# STATISTICS FOR ANALYTICAL CHEMISTRY

# ELLIS HORWOOD SERIES IN ANALYTICAL CHEMISTRY

*Series Editors:* Dr. R. A. CHALMERS and Dr. MARY MASSON
University of Aberdeen

# STATISTICS FOR ANALYTICAL CHEMISTRY

J. C. MILLER, M.A., Ph.D.
Our Lady's Convent School, Loughborough

and

J. N. MILLER, M.A., Ph.D.
Professor of Analytical Chemistry
Loughborough University of Technology

ELLIS HORWOOD LIMITED
Publishers · Chichester

Halsted Press: a division of
JOHN WILEY & SONS
New York · Chichester · Brisbane · Toronto

First published in 1984 by
**ELLIS HORWOOD LIMITED**
Market Cross House, Cooper Street, Chichester, West Sussex, PO19 1EB, England

*The publisher's colophon is reproduced from James Gillison's drawing of the ancient Market Cross, Chichester.*

**Distributors:**

*Australia, New Zealand, South-east Asia:*
Jacaranda-Wiley Ltd., Jacaranda Press,
JOHN WILEY & SONS INC.,
G.P.O. Box 859, Brisbane, Queensland 40001, Australia

*Canada:*
JOHN WILEY & SONS CANADA LIMITED
22 Worcester Road, Rexdale, Ontario, Canada.

*Europe, Africa:*
JOHN WILEY & SONS LIMITED
Baffins Lane, Chichester, West Sussex, England.

*North and South America and the rest of the world:*
Halsted Press: a division of
JOHN WILEY & SONS
605 Third Avenue, New York, N.Y. 10016, U.S.A.


© 1984 J.C. Miller and J.N. Miller/Ellis Horwood Limited

**British Library Cataloguing in Publication Data**
Miller, J.C.
Statistics for analytical chemistry. —
(Ellis Horwood series in analytical chemistry)
1. Chemistry, Analytic — Statistical methods
I. Title  II. Miller, J.N.
519.5'024541   QD75.4.S8

**Library of Congress Card No.** 84-19271

ISBN 0-85312-662-3 (Ellis Horwood Limited — Library Edn.)
ISBN 0-85312-655-0 (Ellis Horwood Limited — Student Edn.)
ISBN 0-470-20128-2 (Halsted Press)

Typeset by Ellis Horwood Limited.
Printed in Great Britain by The Camelot Press, Southampton.

# Table of Contents

# Preface

To add yet another volume to the already numerous texts on statistics might seem to be an unwarranted exercise, yet the fact remains that many highly competent scientists are woefully ignorant of even the most elementary statistical methods. It is even more astonishing that analytical chemists, who practise one of the most quantitative of all sciences, are no more immune than others to this dangerous, but entirely curable, affliction. It is hoped, therefore, that this book will benefit analytical scientists who wish to design and conduct their experiments properly, and extract as much information from the results as they legitimately can. It is intended to be of value to the rapidly growing number of students specializing in analytical chemistry, and to those who use analytical methods routinely in everyday laboratory work.

There are two further and related reasons that have encouraged us to write this book. One is the enormous impact of microelectronics, in the form of microcomputers and hand-held calculators, on statistics: these devices have brought lengthy or difficult statistical procedures within the reach of all practising scientists. The second is the rapid development of new 'chemometric' procedures, including pattern recognition, optimization, numerical filter techniques, simulations and so on, all of them made practicable by improved computing facilities. The last chapter of this book attempts to give the reader at least a flavour of the potential of some of these newer statistical methods. We have not, however, included any computer programs in the book — partly because of the difficulties of presenting programs that would run on all the popular types of microcomputer, and partly because there is a substantial range of suitable and commercially available books and software.

The availability of this tremendous computing power naturally makes it all the more important that the scientist applies statistical methods rationally and correctly. To limit the length of the book, and to emphasize its practical bias, we have made no attempt to describe in detail the theoretical background of the statistical tests described. But we have tried to make it clear to the practising analyst which tests are appropriate to the types of problem likely to be encountered in the laboratory. There are worked examples in the text, and exercises for the reader at the end of each chapter. Many of these are based on the data provided by research papers published in *The Analyst*. We are deeply grateful to Mr. Phil Weston, the Editor, for allowing us thus to make use of his distinguished journal. We also thank our colleagues, friends and family for their forebearance during the preparation of the book; the sources of the statistical tables, individually acknowledged in the appendices; the Series Editor, Dr. Bob Chalmers; and our publishers for their efficient cooperation and advice.

J. C. Miller

J. N. Miller

April 1984

# Glossary of Symbols

$a$ — intercept of regression line
$b$ — gradient of regression line
$c$ — number of columns in two-way ANOVA
$C$ — correction term in two-way ANOVA
$F$ — the ratio of two variances
$F_R$ — ratio of ranges
$h$ — number of samples in one-way ANOVA
$\mu$ — arithmetic mean of a population
$M$ — number of minus signs in Wald–Wolfowitz runs test
$n$ — sample size
$N$ — number of plus signs in Wald–Wolfowitz runs test
$N$ — total number of measurements in two-way ANOVA
$\nu$ — number of degrees of freedom
$P(r)$ — probability of $r$
$Q$ — Dixon's $Q$, used to test for outliers
$r$ — product–moment correlation coefficient
$r$ — number of rows in two-way ANOVA
$\rho$ — Spearmann rank correlation coefficient
$s$ — standard deviation of a sample
$s_{y/x}$ — standard deviation of $y$-residuals
$s_b$ — standard deviation of slope of regression line
$s_a$ — standard deviation of intercept of regression line
$s_{(y/x)_W}$ — standard deviation of $y$-residuals of weighted regression line
$s_{x_0}$ — standard deviation of $x$-value estimated using regression line
$s_B$ — standard deviation of blank
$s_{x_E}$ — standard deviation of extrapolated $x$-value
$s_{x_{0W}}$ — standard deviation of $x$-value estimated by using weighted regression line
$\sigma$ — standard deviation of a population
$\sigma_0^2$ — measurement variance
$\sigma_1^2$ — sampling variance

$t$ — quantity used in the calculation of confidence limits and in significance testing of mean (see Section 2.4)

$T$ — grand total in ANOVA

$T_d$ — quantity used in significance testing of mean (see Section 5.5)

$T_1$ and $T_2$ — test statistics used in the Wilcoxon rank sum test

$w$ — range

$w_i$ — weight given to point on regression line

$\bar{x}$ — arithmetic mean of a sample

$x_0$ — $x$-value estimated by using regression line

$x_E$ — extrapolated $x$-value

$\bar{x}_W$ — arithmetic mean of weighted $x$-values

$\chi^2$ — quantity used to test for goodness-of-fit

$\hat{y}$ — $y$-values predicted by regression line

$\bar{y}_W$ — arithmetic mean of weighted $y$-values

$y_B$ — signal from blank

$z$ — standard normal variable

# 1

# Introduction

## 1.1 ANALYTICAL PROBLEMS

An analytical chemist may be presented with two types of problem. Sometimes he is asked to provide only a qualitative answer. For example, the presence of boron in distilled water is very damaging in the manufacture of microelectronic components — "Does this distilled water sample contain any boron?" Again, the comparison of soil samples is a common problem in forensic science — "Could these two soil samples have come from the same site?" In other cases the problems posed are quantitative ones. "How much albumin is there in this sample of blood serum?" "How much lead in this sample of tap-water?" "This steel sample contains small quantities of chromium, tungsten and manganese — how much of each?" These are typical examples of single-component or multiple-component quantitative analyses.

Modern analytical chemistry is overwhelmingly a **quantitative** science. It is apparent that in many cases a quantitative answer will be much more valuable than a qualitative one; it may be useful for an analyst to be able to say that he has detected some boron in a distilled water sample, but it is much more useful for him to be able to say *how much* boron is present. The person who requested the analysis could, armed with this quantitative answer, judge whether or not the level of boron was of concern, consider how it might be reduced, and so on: but if he knew only that *some* boron was present it would be hard for him to judge the significance of the result. In other cases, it is only a quantitative result that has any value at all. For example, virtually all samples of (human) blood serum contain albumin; the only question is, how much?

It is important to note that even where a qualitative answer is required, quantitative methods are often used to obtain it. This point is illustrated with the aid of the examples given at the beginning of this section. In reality, an analyst would never simply say "I can/cannot detect boron in this water sample". He would use a quantitative method capable of detecting boron at, say, levels of 1 $\mu$g/ml. If his test gave a negative result, it would then be described in the form "This sample contains less than 1 $\mu$g/ml boron". If the test gave a positive result the sample would be reported to contain at least 1 $\mu$g/ml boron. Much more complex quantitative approaches might be used to compare two soil samples. For example, the samples might be subjected to a particle-size analysis, in which the soil particles are classified according to size, in a number (say ten) of size ranges, and the fraction of sample in each range is determined. Each sample would then be characterized by these ten pieces of data. Quite complex procedures (see Chapter 6) can then be used to provide a quantitative assessment of their similarity, and an estimate of the probability of the samples having a common origin can be given if comparable data from soil samples *known* to match are available.

## 1.2 ERRORS IN QUANTITATIVE ANALYSIS

Once we accept that quantitative studies will play a dominant role in any analytical laboratory, we must accept also that the errors that occur in such studies are of supreme importance. Our guiding principle will be that *no quantitative results are of any value unless they are accompanied by some estimate of the errors in them.* This principle naturally applies not only to analytical chemistry but to any field of study in which numerical experimental results are obtained. We can readily examine a number of simple examples which not only illustrate the principle but also foreshadow the types of statistical problem we shall meet and solve in subsequent chapters.

Suppose a chemist synthesizes an analytical reagent which he believes to be entirely new. He studies it by using a spectrometric method and the compound gives a value of 104 (normally, most of our results will be cited in carefully-chosen units, but in this hypothetical example purely arbitrary units can be used). On checking the reference books, the chemist finds that no compound previously discovered has yielded a value of more than 100 when studied by the same method in the same experimental conditions. The question thus naturally arises, has our chemist really discovered a new compound? The answer to this question lies of course in the degree of reliance that we can place on that experimental value of 104. What errors are associated with it? If further study indicates that the result is correct to within 2 (arbitrary) units, i.e. the true value probably lies in the range 104 ± 2, then a new material has probably been characterized. If, however, investigations show that error may amount to 10 units (i.e. 104 ± 10), then it is quite likely that the true value is actually less than 100, in which case a new discovery is far from certain. In other words, a knowledge of the experimental errors is crucial (in this case as in every other) to the proper interpretation of the results. In statistical terms this example would involve the comparison of the experimental result with an assumed or reference value: this topic is studied in detail in Chapter 3.

A more common situation is that of the analyst who performs several replicate determinations in the course of a single analysis. (The value and significance of such replicates is discussed in detail in the next chapter.) Suppose an analyst performs a titrimetric experiment four times and obtains values of 24.69, 24.73, 24.77 and 25.39 ml. The first point to notice is that titration values are reported to the nearest 0.01 ml; this point is also discussed further in Chapter 2. It is also immediately apparent that all four values are different, because of the errors inherent in the measurements, and that the fourth value (25.39 ml) is substantially different from the other three. The question that arises here is — can this fourth value be safely rejected, so that (for example) the mean value is reported as 24.73 ml? In statistical terms, is the value 25.39 ml an outlier? The important topic of outlier rejection is discussed in detail in Chapter 3.

Another frequent problem involves the comparison of two (or more) sets of results. Suppose that an analyst measures the vanadium content of a steel sample by two separate methods. With the first method he obtains an average value of 1.04%, with an estimated error of 0.07%; using the second method he obtains an average value of 0.95% and an error of 0.04%. Several questions arise from the comparison of these results. Are the two average values significantly different, or are they indistinguishable within the limits of experimental error? Is one method significantly less error-prone than the other? Which of the mean values is actually closer to the truth? Again, Chapter 3 discusses these and related questions.

To conclude this section we note that many analyses are based on graphical methods. Instead of making repeated measurements on the same sample, we perform a series of measurements on a small group of standards which have known concentrations covering a considerable range. In this way we set up a calibration curve that can be used to estimate the concentrations of test samples studied by using the same procedure. In practice, of course, all the measurements (those utilizing the standards and those on the test samples) will be subject to errors. It is necessary, for example, to assess the errors involved in drawing the calibration curve; to estimate the error in the concentration of a single sample determined by using the curve; and to estimate the limit of detection of the method, i.e. the smallest quantity of analyte that can be detected with a particular degree of confidence. These procedures, which are especially commonplace in instrumental analysis, are described in Chapter 4.

These examples represent only a fraction of the possible problems arising from the occurrence of experimental errors in quantitative analysis. As we have seen, however, the problems have to be solved if the quantitative data are to have any real meaning. It is thus apparent that we must study the various types of error in more detail.

## 1.3  TYPES OF ERROR

Experimental scientists make a fundamental distinction between three types of error. These are known as **gross, random,** and **systematic** errors. Gross errors are readily described: they may be defined as errors which are so serious that there is no real alternative to abandoning the experiment and making a completely fresh start. Examples would include a complete instrument breakdown, accidentally dropping or discarding a crucial sample, or discovering during the course of the experiment that a supposedly pure reagent was in fact badly contaminated. Such errors (which occur occasionally even in the best-regulated laboratories!) are normally very easily recognized. In our discussion we thus have only to distinguish carefully between **random** and **systematic** errors.

We can best make this distinction by careful study of a real experimental situation. Four students (A-D) each perform an analysis in which *exactly* 10.00 ml of *exactly* 0.1*M*\* sodium hydroxide is titrated with *exactly* 0.1*M* hydrochloric acid. Each student performs five replicate titrations, with the results shown in Table 1.1.

**Table 1.1** – Random and systematic errors

| Student | Results (ml) | Comment |
|---------|--------------|---------|
| A | 10.08<br>10.11<br>10.09<br>10.10<br>10.12 | Precise but inaccurate |
| B | 9.88<br>10.14<br>10.02<br>9.80<br>10.21 | Accurate but imprecise |
| C | 10.19<br>9.79<br>9.69<br>10.05<br>9.78 | Inaccurate and imprecise |
| D | 10.04<br>9.98<br>10.02<br>9.97<br>10.04 | Accurate and precise |

The results obtained by student A have two important characteristics. First, they are all very close; all the results lie between 10.08 and 10.12 ml. In everday terms we would say that the results are highly reproducible; we discuss this term in more detail below. The second distinctive feature of the results is that they

---

\*The symbol 1*M* stands for 1 mole of the specified material per litre of solution.

are *all too high:* in this experiment (somewhat unusually) we know the correct answer in advance — it should be 10.00 ml. It is evident that two entirely separate types of error have occurred within this student's experiment. First, there are **random** errors — *these cause the individual results to fall on both sides of the average value* (10.10 ml in this case). Statisticians say that **random** errors affect the **precision**, or **reproducibility**, of an experiment. In the case of student A it is clear that the random errors are small, so we say that the results are **precise.** In addition, however, there are **systematic errors** — *these cause all the results to be in error in the same sense* (in this case they are all too high). Systematic errors affect **accuracy**, i.e. *proximity to the true value.* In many experiments the random and systematic errors have quite distinct origins in terms of experimental technique and equipment. Before we examine the causes of the errors in this experiment, however, we can discuss briefly the results obtained by students B-D. Student B has obtained results which are in direct contrast to those of student A. The average of the five results (10.01 ml) is very close to the true value, so we could characterize the data as accurate — without substantial systematic error. The spread of the results is very large, however, indicating poor precision and the presence of substantial random errors. Comparison of these results with those obtained by student A shows clearly that random and systematic errors can occur independently of one another. This conclusion is reinforced by the data of students C and D. Student C's work is neither precise (range 9.69–10.19 ml) nor accurate (average 9.90 ml). Student D has achieved both precise (range 9.97–10.04 ml) and accurate (average 10.01 ml) results. The distinction between random and systematic errors, and precision and accuracy respectively, is summarized in Fig. 1.1.

The apparently simple experiments described in the previous paragraphs merit several additional comments. It is especially noteworthy that the words **precision** and **accuracy** have entirely distinct meanings in the theory of errors, whereas they are used interchangeably in everyday English. Several dictionaries, for example, cite accuracy as a definition of precision (and vice versa!). This unfortunate confusion undoubtedly makes it more difficult to remember the essential difference between random and systematic errors, to which the words precision and accuracy respectively refer. Two further points of nomenclature may be mentioned here. Some texts speak of 'determinate' and 'indeterminate' errors, meaning the same as systematic and random errors respectively. Also, although we earlier used the word 'reproducibility' as an approximate definition of precision, modern convention makes a careful distinction between **reproducibility** and **repeatability**.

We can illustrate this distinction by an extension of our previous experiment. In the normal way student A (for example) would do the five replicate titrations in rapid succession; very probably the whole exercise would not take more than an hour or so. The same set of solutions and the same glassware would be used throughout, the same preparation of indicator would be added to each titration

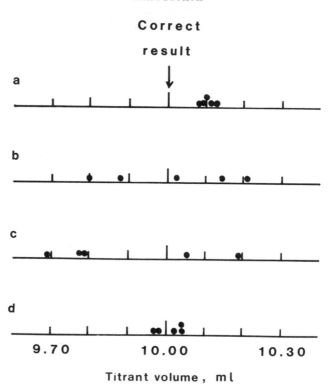

Fig. 1.1 – Accuracy and precision – graphical representation of the data in Table 1.1. In (a) the data are precise but inaccurate, in (b) accurate but imprecise, in (c) inaccurate and imprecise, and in (d) accurate and precise.

flask, and the temperature, humidity and other laboratory conditons would remain much the same. In such circumstances the precision measured would be the *within-run* precision: this is called the **repeatability**. Suppose, however, that for some reason the same student did the titrations on five different occasions. In such circumstances he or she would probably use different pieces of glassware and different batches of indicator; the laboratory conditions might also change from time to time. So it would not be surprising to find a greater spread of the results in this case. This set of data would reflect the *between-run* precision of the method, and it is this to which the term **reproducibility** should be applied. Table 1.2 summarizes the definitions of the terms used thus far and the relationships between them.

One further lesson may be deduced from the titration experiment. It is easy to appreciate that the data obtained by student C are unacceptable, and that those of student D are the most acceptable. Sometimes, however, it may happen that two methods are available for a particular analysis, of which one is believed

**Table 1.2** – Summary of definitions

| Types of error | |
|---|---|
| Random | Systematic |
| Affect precision | Affect accuracy |
| Within-run precision is **repeatability** | Proximity to the truth |
| Between-run precision is **reproducibility** | |
| Also known as indeterminate errors | Also known as determinate errors |

to be precise but inaccurate, while the other is accurate but imprecise. In other words we may have to choose between the types of results obtained by students A and B respectively. Which type of result is preferable? It is impossible to give a dogmatic answer to this question, if only because the choice of analytical method will often in practice be based on the cost, ease of automation, speed of analysis, and other factors falling outside our elementary evaluation. It is nonetheless important to realize that a method which is substantially free from systematic errors may still, if it is very imprecise, give an average value that is (by chance) considerably removed from the correct value. On the other hand a method that is precise but inaccurate (cf. student A) can be converted into one that is both precise *and* accurate (cf. student D) if the systematic errors can be discovered and hence removed. There will also be cases where, because the measurements being attempted are completely new, no check for systematic errors is feasible. Random errors can never be eliminated, though by careful technique we can minimize them, and by making repeated measurements we can measure them and evaluate their significance. Systematic errors can in many cases be removed by careful checks on our experimental technique and equipment. This additional crucial distinction between the two major types of error is further explored in the next section.

## 1.4 RANDOM AND SYSTEMATIC ERRORS IN TITRIMETRIC ANALYSIS

The example of the students' titrimetric experiment showed clearly that random and systematic errors can occur independently of one another and thus presumably arise at different stages of the experiment. Since titrimetry is a relatively simple yet still widely adopted procedure it is of interest and value to examine it in detail in this context. A complete titrimetric analysis can be regarded as having the following steps (some of which were omitted from the example in the preceding section in the interest of simplicity).

(1) Making up a standard solution of one of the reactants. This involves (a) weighing a weighing bottle or similar vessel containing some solid material, (b) transferring the solid material to a standard flask and weighing the bottle again to obtain by subtraction the weight of solid transferred, and (c) filling the flask up to the mark with water (assuming that an aqueous titration is contemplated).

(2) Transferring an aliquot of the standard material to a titration flask with the aid of a pipette. This involves (a) filling the pipette to the appropriate mark, and (b) draining it in a specified manner into the titration flask.

(3) Titrating the liquid in the flask with a solution of the other reactant added from a burette. This involves (a) filling the burette and allowing the liquid in it to drain until the meniscus is at a constant level, (b) adding a few drops of indicator solution to the titration flask, (c) reading the initial burette volume, (d) adding liquid to the titration flask from the burette a little at a time until the end-point is adjudged to have been reached, and (e) measuring the final level of the liquid in the burette.

Even an elementary analysis of this type thus involves some ten separate steps, the last seven of which are normally, as we have seen, repeated several times. In principle, we should examine each step to evaluate the random and systematic errors which might occur. In practice, it is simpler to examine separately those stages which utilize weighings [steps 1(a) and 1(b)], and the remaining stages involving the use of volumetric equipment. The reader should notice that it is not the intention to give detailed descriptions of the experimental techniques used in the various stages. Similarly, methods for calibrating weights, glassware etc. will not be given. (These topics are covered in adequate detail by the books listed in the bibliography at the end of the chapter). Important amongst the contributions to the errors are the tolerances of the weights used in the gravimetric steps, and of the volumetric glassware. Standard specifications for these tolerances are issued by such bodies as the British Standards Institution (BSI) and the American Society for Testing and Materials (ASTM). Some typical tolerance values are given in Table 1.3. The tolerance of a top-quality 100-g weight can be as little as ±0.25 mg, although for a weight used in routine work the tolerance would be up to four times as large. Similarly the tolerance for a grade A 250-ml standard flask is ±0.12 ml: grade B glassware generally has tolerances twice as large as grade A glassware. If a weight or a piece of glassware is within the tolerance limits, but not of exactly the correct weight or volume, a systematic error will arise. Thus, if the standard flask actually has a volume of 249.95 ml, this error will be reflected in the results of all the experiments based on use of that flask. Repetition of the experiment will not reveal the error; in each replicate the volume will be assumed to be 250.00 ml when in fact it is less than this. If, however, the results of an experiment using this flask are compared with the results of several other experiments (e.g. in other laboratories) using

**Table 1.3** – Examples of tolerances of weights and glassware

| Item | Tolerance |
|---|---|
| Weights (Class 1) | |
| 100 g | ±0.25 mg |
| 1 g | 0.034 mg |
| 10 mg | 0.010 mg |
| | |
| Weights (Class 3) | |
| 100 g | ±1.0 mg |
| 1 g | 0.10 mg |
| 10 mg | 0.030 mg |
| | |
| Glassware (Grade A) | |
| 50 ml measuring cylinder | 0.25 ml |
| 250 ml standard flask | 0.12 ml |
| 25 ml pipette | 0.03 ml |
| 50 ml burette | 0.05 ml |

(Data from ASTM Standard Specifications E617–78 and E694–79; see bibliography).

other flasks, then the fact that all the flasks will have slightly different volumes will contribute to the random variation: it will be recalled that between-run random errors are said to affect the reproducibility of the results.

Weighing procedures are normally associated with very small **random** errors. In routine laboratory work a four-place balance is commonly used, and the random error involved should not be greater than ±0.0001–0.0002 g (the next chapter describes in detail the statistical terms used to describe random errors). Given that the quantity being weighed is normally about 1 g or more, it is evident that the random error, expressed as a percentage of the weight involved, is not more than 0.02%. A good standard material for volumetric analysis should (amongst other characteristics) have as high a formula weight as possible, so as to minimize these random errors when a solution of a given molarity is being made up. In some analyses microbalances are used to weigh quantities of a few milligrams — but the weighing errors involved are likely to be a few micrograms.

**Systematic** errors in weighings can be appreciable, and have a number of well-established sources. These include adsorption of moisture on the surface of the weighing vessel; failure to allow heated vessels to cool to the same temperature as the balance before weighing (this error is especially commonplace in gravimetry when crucibles are weighed); corroded or dust-contaminated weights; and the buoyancy effect of the atmosphere, which acts to a different extent on objects of equal mass but different density. For the most accurate work, weights must be calibrated against standards furnished by statutory bodies and standards laboratories. This calibration can be very accurate indeed, e.g. to ±0.01 mg for weights in the range 1–10 g. The buoyancy effect can be substantial. For example Skoog and West (see bibliography) show that a sample of an organic liquid of density 0.92 g/ml which weighs 1.2100 g in air would weigh 1.2114 g in vacuo, a difference of more than 0.1%. Apart from the use of calibration procedures, which are further discussed in the next section, some simple experimental precautions can be taken to minimize these systematic errors. The most important of these is to perform all weighings **by difference**, i.e. a weighing bottle is weighed first with the sample contained in it, and secondly with the sample removed. The **difference** between the two weights gives the weight of the sample, and also cancels the systematic errors arising from (for example) the moisture and other contaminants on the surface of the bottle. (See also Section 2.8). *If such precautions are taken,* the errors in the weighing steps will be small, and it is probable that in most volumetric experiments weighing errors will be negligible compared with the errors arising from the use of volumetric equipment. Indeed, gravimetric methods are generally used for the calibration of volumetric glassware, by weighing (in standard conditions) the water that it contains; this procedure is obviously only valid because gravimetric errors are negligible compared with volumetric ones.

The random errors in volumetric procedures arise in the use of volumetric glassware. In filling a 250-ml standard flask to the mark, the error (i.e. the distance between the meniscus and the mark) might be about ±0.03 cm in a flask neck of diameter ca. 1.5 cm. This corresponds to a volume error of about 0.05 ml – only 0.02% of the total volume of the flask. Similarly the random error in filling a 25-ml transfer pipette should not exceed 0.03 cm in a stem of diameter 0.5 cm; this gives a volume error of ca. 0.006 ml, 0.024% of the total volume. The error in reading a burette (of the conventional variety graduated in 0.1-ml divisions) is perhaps 0.01–0.02 ml. Each titration involves two such readings (the errors of which are *not* simply additive – see Chapter 2); if the titration volume is ca. 25 ml, the percentage error is again very small. It is clear that the experimental conditions should be arranged so that the volume of titrant used is not too small (say not less than 10 ml), otherwise the errors will become appreciable. (This precaution is analogous to the use of a standard compound of high formula weight in a gravimetric step). Even though a volumetric analysis involves several steps, in each of which a piece of volumetric glassware is used, it is apparent that the random errors should be small if the experiments are performed with care. In practice a good volumetric analysis should have a relative standard deviation (see Chapter 2) of not more than 0.1%. Until recently, such precision was seldom attainable in instrumental analysis methods (and is still not commonplace), and this constituted one of the main advantages of classical analysis techniques. It should be remembered that in skilled hands, with all precautions taken, classical methods can give results with relative standard deviations of 0.01%.

Volumetric procedures incorporate several important sources of systematic error. Chief amongst these are the drainage errors in the use of volumetric glassware, the effects of temperature on the volume of liquids and glassware, and 'indicator errors'. Perhaps the commonest error in routine volumetric analysis is to fail to allow enough time for a pipette to drain properly, or a meniscus level in a burette to stabilize. It is also worth remembering that pipettes are of two types, those which are emptied by drainage, and blow-out pipettes from which the last remaining liquid must be forcibly expelled. To confuse the two types, for example to blow out a drainage pipette, would certainly count as a gross error! Drainage errors have a systematic as well as a random effect — the volume delivered is invariably less than it should be. The temperature at which an experiment is performed has two effects. Volumetric equipment is conventionally calibrated at 20°C, but the temperature in an analytical laboratory may easily be several degrees different from this, and many experiments, for example in biochemical analysis, are carried out in "cold rooms" at ca. 4°C. Secondly, the temperature affects the volumes of liquids. The coefficient of expansion for dilute aqueous solutions is about 0.025% per degree, whereas a soda-glass vessel will change in volume by about 0.003% per degree and a borosilicate glass vessel by about 0.001% per degree. It is evident that the changes in the volumes of glassware will only be important in work of the highest accuracy, and even then only if the temperature is very different from 20°C. Furthermore, the effects of the expansion of the solutions will be largely self-cancelling if all the solutions are maintained at the same temperature.

Indicator errors can be quite substantial — perhaps larger than the random errors in a typical titrimetric analysis. For example, in the titration of 0.1$M$ hydrochloric acid with 0.1$M$ sodium hydroxide, we expect the end-point to correspond to a pH of 7. In practice, however, we estimate this end-point by the use of an indicator such as Methyl Orange. Separate experiments show that this substance changes colour over the pH range ca. 3–4. If, therefore, the titration is performed by adding alkali to acid, the indicator will yield an apparent end-point when the pH is ca. 3.5, i.e. just before the true end-point. The systematic error involved here is likely to be as much as 0.2%. If the titration is performed in reverse, i.e. by adding acid to alkali, the end-point indicated by the Methyl Orange will actually be a little beyond the true end-point, because of the need to achieve an acid pH to effect the colour change. Thus the systematic indicator error can be either positive or negative according to the design of the experiment; with a given procedure it will be either always positive or always negative. In either case the error can be evaluated and corrected by performing a **blank** experiment. This point is considered in more detail in the next section. A seldom mentioned error is the "last drop" error. It is not known *how much* of the last drop added is actually needed to reach the end-point. Since either less or more than half the volume of the drop must be needed, this error can be halved by subtracting half the volume of a drop.

In any analytical procedure, classical or instrumental, it is possible to consider and estimate the sources of random and systematic error arising at each separate stage of the experiment. It is very desirable for the analyst to do this, as it may enable him to avoid major sources of error by careful experimental design (see Section 1.6). It is worth noting, however, that titrimetric analyses are rather unusual in that they involve no single step having an error which is far greater than the errors in the other steps. In many analyses the overall error is in practice dominated by the error in a single step: this point is further discussed in the next chapter.

## 1.5 HANDLING SYSTEMATIC ERRORS

Much of the remainder of this book will deal with the evaluation of random errors, which can be studied by a wide range of statistical methods. In most cases we shall assume for convenience that systematic errors are absent (though methods which test for the occurrence of systematic errors will be described). It is thus necessary at this stage to discuss systematic errors in more detail, how they arise, and how they may be combated. The example of the titrimetric analysis given in Section 1.3 shows clearly that systematic errors cause the mean value of a set of replicate measurements to deviate from the true value. It follows that (a) in contrast to random errors, systematic errors cannot be revealed merely by making repeated measurements, and that (b) unless the true result of an analysis is known in advance − a most unlikely situation! − very large systematic errors might occur, but go entirely undetected unless suitable precautions are taken. In other words, it is all too easy completely to overlook substantial sources of systematic error. A small number of examples will clarify both the possible problems and their solutions.

In recent years much interest has been shown in the levels of transition metals in biological samples such as blood serum. Many determinations have been made of the levels of (for example) chromium in serum − with startling results. Different workers, all studying pooled serum samples from healthy subjects, have obtained chromium concentrations varying from <1 to ca. 200 ng/ml! In general, the lower results have been obtained more recently, and it has gradually become apparent that the earlier, higher values were due at least in part to contamination of the samples by chromium from stainless-steel syringes, tube caps, and so on. The determination of traces of chromium, for example by atomic-absorption spectrometry, is in principle relatively straightforward, and no doubt each group of workers achieved results which seemed satisfactory in terms of precision, but in a number of cases the large systematic error introduced by the contamination was entirely overlooked. Methodological systematic errors of this kind are extremely common − incomplete washing of a precipitate in gravimetric analysis, and the indicator error in volumetric analysis (see Section 1.4) are further well-known examples.

Another class of systematic error that occurs widely arises when false assumptions are made about the accuracy of an analytical instrument. Experienced analysts know only too well that the monochromators in spectrometers gradually go out of adjustment, so that errors of several nanometres in wavelength settings are not uncommon, yet many photometric analyses are undertaken without appropriate checks being made. Very simple devices such as volumetric glassware, stop-watches, pH-meters and thermometers can all show substantial systematic errors, but many laboratory workers regularly use these instruments as though they are always perfectly accurate. Moreover, the increasing availability of instruments controlled by microprocessors or microcomputers has reduced to a minimum the number of operations and the level of skill required of their operators. In these circumstances the temptation to regard the instrument's results as beyond reproach is overwhelming — yet such instruments (unless they are "intelligent" enough to be self-calibrating — see below) are still subject to systematic errors.

Systematic errors arise not only from procedures or from apparatus; they can also arise from human bias. Some chemists suffer from a degree of colour-blindness, astigmatism, or other defects of eyesight which might introduce errors into their readings of instruments and other observations. A number of authors have reported various types of number bias, for example a tendency to favour even over odd numbers, or 0 and 5 over other digits, in the reporting of results. It is thus apparent that systematic errors of several kinds are a constant, and often hidden, risk for the analyst, so the most careful steps to minimize them must be considered. Several different approaches to this problem are available, and any or all of them should be considered in each analytical procedure.

The first precautions should be taken before any experimental work is begun. The analyst should consider carefully each stage of the experiment he is about to perform, the apparatus to be used and the sampling and analytical procedures to be used. He should try to identify at this earliest stage the likely sources of systematic error, such as the instrument functions that need calibrating, the steps of the analytical procedure where errors are most likely to occur, and should also consider the checks he can subsequently make for systematic errors. Foresight of this kind can be immensely valuable (we shall see in the next section that similar advance attention should be given to the sources of random error) and is normally well worth the time invested. For example, a little thinking of this kind might well have revealed the probability of contamination occurring in the chromium experiment described above.

The second line of defence against systematic errors lies in the design of the experiment at every stage. We have already seen (Section 1.4) that weighing by difference can remove some systematic gravimetric errors: these errors can be assumed to occur to the same extent in both weighings, so the subtraction process eliminates them. A further example of thoughtful experimental planning is provided by the spectrometer wavelength error described above. If the concentration of a sample is to be determined by absorption spectrometry, two procedures are possible. In the first, the sample is studied in a 10-mm spectrometer cell at a single wavelength, say 400 nm, and the concentration of the test component is determined from the well-known equation $A = \epsilon c l$ [where A, $\epsilon$, c and l are respectively the measured absorbance, the molar absorptivity (units l.mole$^{-1}$, cm$^{-1}$), the molar concentration of the test component, and the path-length (cm) of the spectrometer cell]. Several systematic errors can arise here: the wavelength might, as already discussed, be (say) 405 nm rather than 400 nm, thus rendering the reference value of $\epsilon$ inappropriate; the reference value of $\epsilon$ might in any case be wrong; the absorbance scale of the spectrometer might also exhibit a systematic error; the path-length of the cell might not be exactly 10 mm. Alternatively, the analyst might take a series of solutions of the test substance of *known* concentration, and measure the absorbance of each at 400 nm. The results would then be used to construct a calibration graph for the analysis of the test sample under exactly the same experimental conditions. (This important approach to instrumental analysis is described in detail in Chapter 4.) When this second method is used, the value of $\epsilon$ is not required, and the errors due to wavelength shifts, absorbance errors and path-length inaccuracies should cancel out, as they occur equally in the calibration and test experiments. Provided only that the conditions are indeed equivalent for the test and calibration samples (for example, we are assuming that the wavelength and absorbance scales do not alter *during the experiment*) all the major sources of systematic error are in principle eliminated.

The final and perhaps most formidable protection against systematic errors is the use of standard reference materials and methods. Before the experiment is started each piece of apparatus is calibrated by an appropriate procedure. We have already seen that volumetric equipment can be calibrated by the use of gravimetric methods. Similarly, spectrometer wavelength scales can be calibrated with the aid of standard light sources which have narrow emission lines at well-established wavelengths, and spectrometer absorbance scales calibrated with standard solid or liquid filters. In analogous fashion, most pieces of equipment can be calibrated so that their systematic errors are known in advance. The importance of this area of chemistry (and other experimental sciences) is reflected in the extensive work of bodies such as the National Physical Laboratory (in the U.K.), the National Bureau of Standards (in the U.S.A.) and similar organizations elsewhere. Whole volumes have been written on the standardization of particular types of equipment (see bibliography), and a number of commercial organizations specialize in the sale of standard reference materials. If systematic errors are occurring during chemical processes or as a result of using impure reagents, rather than in the equipment, an alternative form of comparison must be used, i.e. the determination must be repeated by use of an entirely independent procedure. If two (or more) chemically and physically unrelated methods are used to perform one analysis, and if they consistently yield results showing only random differences, it is a reasonable presumption that no significant systematic errors are present. It is clear that for this approach to be valid, each step of the two experiments has to be independent. Thus in the case of the serum chromium determinations, it would not be sufficient to replace the atomic-absorption spectrometry step by, for example, a colorimetric method or by plasma spectrometry: the systematic errors would only be revealed by altering the sampling methods also, e.g. by minimizing or eliminating the use of stainless-steel syringes. A further important point is that comparisons must be made over the whole of the concentration range for which an analytical procedure is to be used. For example, the widely-used Bromocresol Green dye-binding method for the determination of albumin in serum correlates well with alternative methods (e.g. immunological methods) at normal or high levels of albumin, but when the albumin levels are abnormally low (these are inevitably the cases of most interest!) the agreement between the two methods is sometimes poor, the dye-binding method giving consistently (and erroneously) higher albumin concentrations.

The mathematical methods used in these comparisons are described in detail in Chapters 3 and 4. The examples at the end of this chapter will provide the reader with some practice in identifying possible systematic errors and attempting their elimination.

## 1.6 PLANNING AND DESIGN OF EXPERIMENTS

Many chemists regard statistical tests as methods to be used only to assess the results of completed experiments. While this is indeed a crucial area of application of statistics — we have seen that no quantitative result is worth anything if it is not accompanied by an estimate of its errors — it is also necessary to be aware of the importance of statistical concepts in the planning and design of experiments. In the previous section the value of trying to predict systematic errors in advance, thereby permitting the analyst to lay plans for countering them, was emphasized. The same considerations apply to random errors. As will be seen in Chapter 2, the combination of the random errors of the individual parts of an experiment to give an overall error requires the use of simple statistical formulae. In practice, the overall error is often dominated by the error in just one stage of the experiment, other smaller errors having negligible effects when all the errors are combined correctly. Again it is obviously desirable to try to identify *before the experiment begins* where this single dominant error is likely to arise, and then to try to minimize it. (Although random errors can never be eliminated, they can certainly be minimized by particular attention to experimental techniques: improving the precision of a spectrometric experiment by using a constant-temperature sample cell would be a simple instance of such a precaution.) For both random and systematic errors, therefore, the moral is clear; every effort must be made to identify the serious sources of error before the experiment starts, so that the experiment can be designed to minimize such errors.

We must also give some preliminary consideration to a further, and more subtle, aspect of experimental design. In many analyses, one or more of the desirable features of the method (for example sensitivity, selectivity, sampling rate, low cost etc.) will be found to depend on a number of experimental factors. We shall wish to design the experiment so that the optimum combination of these factors is used, thereby obtaining the best sensitivity, selectivity etc. Although *some* preliminary experiments or prior knowledge will be necessary, optimization should be performed (in the interests of conserving resources and time) before the method is put into routine or widespread use.

The complexity of optimization procedures can be illustrated with the aid of an example. In enzymatic analyses, the concentration of the analyte is determined from observations of the rate of an enzyme-catalysed reaction. The analyte is often the substrate, i.e. the compound that is changed in the reaction catalysed by the enzyme. Let us assume that we are seeking the maximum reaction rate in a particular experiment, and that the rate in practice depends upon (amongst other factors) the pH of the reaction mixture, the temperature, and the concentration of the enzyme. How are the optimum conditions to be found? It is easy to identify one possible approach. The analyst could perform a series of experiments, in each of which the enzyme concentration and the temperature are kept constant but the pH is varied. In each case the rate of the enzyme-catalysed reaction would be determined and an optimum pH value would thus be obtained — let us say that it is 7.5. A second series of reaction-rate experiments could then be performed, with the pH maintained at 7.5, the enzyme concentration again fixed, but the temperature varied. An optimum temperature would thus be found — say 40°C. Finally a series of experiments at pH 7.5 and 40°C but with various enzyme concentrations would indicate the optimum enzyme level. This approach to the optimization of the experiment is clearly tedious. It has the further and more fundamental objection that it assumes that the variables or factors (pH, temperature, enzyme concentration) affect the reaction rate in an independent fashion. This might not be true. For example we have taken it that, at pH 7.5, the optimum temperature is 40°C. At a different pH, however, the optimum temperature might not be 40°C. In other words these variables might affect the reaction rate in an interactive fashion, and the conditions established in the series of experiments just described might not actually be the optimum ones. Had the first series of experiments been started in different conditions, a different set of optimum values might have been obtained.

It will be apparent even from this simple example that experimental optimization can be a formidable problem. In practice, of course, the optimization might involve more than three factors, leading to further complexity. This very important aspect of statistics as applied to analytical chemistry is considered in more detail in Chapter 6.

## 1.7 CALCULATORS AND COMPUTERS IN STATISTICAL CALCULATIONS

No chemist can be unaware of the astonishing developments that have recently taken place in the realm of microelectronics. These advances have made possible the construction of devices which enormously simplify statistical calculations. The rapid growth of 'chemometrics' — the application of mathematical methods to the solution of chemical problems of all types — is due to the ease with which large quantities of data can be handled, and complex calculations done, with calculators and computers.

These devices are available to the analytical chemist at several levels of complexity and cost, as follows.

(1) Hand-held calculators are extremely cheap, very reliable, and capable of performing many routine statistical calculations with a minimal number of key-strokes. In many calculators, preprogrammed functions will allow calculations of mean and standard deviation (see Chapter 2) and correlation and linear regression (see Chapter 4). Other calculators can be fitted with plug-in cards, or programmed by the user, to perform additional calculations such as confidence limits (see Chapter 2), significance tests (see Chapter 3) and non-linear regression (Chapter 4). For many applications in laboratories performing analytical research or routine analyses, calculators of these types will be more than adequate. Their disadvantages lie in their inability to handle very large quantities of data, and the problems arising from the limited number of significant figures incorporated into their algorithms (see Chapter 2).

(2) Microcomputers, sometimes costing little more than calculators, are now to be found in many analytical laboratories. They have to be programmed to perform even the simplest statistical calculations, but can handle quite large quantities of data, and normally use more significant figures than do calculators. Perhaps their greatest value is that they can easily be interfaced to almost any type of laboratory electronic equipment. The microcomputer can then be used to control the instrument's operation, collect the resulting data, and process and record them very rapidly. Microcomputers can also be converted into simple word-processors (this book was written with the help of such a system!), and this facility greatly aids the production of reports on laboratory work. By use of networking systems, several computers can be linked to a central data store, such as a floppy- or hard-disc system. It is possible to interface several instruments to a single computer, though only one instrument can normally be controlled at a time. Similarly the microcomputer can be used as a 'stand-alone' device even when it is interfaced to an instrument, though the two functions will not be available simultaneously.

(3) Many laboratory instruments incorporate their own microprocessors. These systems are normally designed specifically to control the single instrument in question by means of preprogrammed functions. They thus lack the flexibility of microcomputer systems, but may be capable of highly

complex calculations and control operations. A recent advance in this area is the development of so-called 'intelligent' instruments, in which the controlling microprocessor is capable of diagnosing instrument malfunctions and applying the appropriate corrections, or of optimizing a particular analytical procedure. Such developments seem certain to continue, greatly increasing the efficiency of analytical laboratories.

(4) Larger computers, e.g. the so-called mini- and main-frame computers, are now used for the control of only very large and complex instruments. More often they are used in stand-alone mode to handle large quantities of data and/or very complex calculations, for example those requiring numerous iterations. The importance of such devices in many analytical laboratories has declined with the advent of microcomputers: certainly they will only rarely be needed to execute statistical calculations.

It is most important for the analytical chemist to remember that the availability of all these data-handling facilities increases rather than decreases the need for a sound knowledge of the principles underlying statistical calculations. A computer or calculator will rapidly perform such calculations, *whatever the data inserted.* For example, *any* set of $x$ and $y$ values will be fitted to a straight line by a 'least squares' program, even in cases where visual inspection would show at once that such a program was entirely inappropriate (see Chapter 4). The computer or calculator will give no guidance as to which statistical test should be used; it will blindly — but efficiently — perform the calculation the user requests. The analyst must thus use his knowledge of statistics, and his common sense, to ensure that the appropriate calculation is performed. In instruments where a built-in microprocessor is used to calculate analytical results, it is especially important for the user to know the basis of the calculation. This has recently become a matter of some controversy: some instrument manufacturers are understandably reluctant to reveal the details or even the underlying principles of computer programs that are unique to individual instruments and have cost a good deal to develop.

In this book no attempt will be made to provide complete listings of computer programs for statistical calculations. Estimations of means, standard deviations, confidence limits, linear regression parameters, and some of the simple statistical tests, are readily performed on preprogrammed or programmable hand-held calculators. There is a great deal of commercial software available for performing such calculations (often more rapidly, and/or with large quantities of data) on microcomputers interfaced to analytical instruments. More sophisticated statistical tests, non-parametric tests (cf. Chapter 5), and procedures involving pattern recognition, optimization and other advanced methods are also available in the form of software for the popular microcomputers.

## BIBLIOGRAPHY

1. Many conventional textbooks of analytical chemistry include sections or chapters on the statistical treatment of experimental results. These sections are very variable in their coverage, but a very good example is Chapter 3 of *Fundamentals of Analytical Chemistry* by D.A. Skoog and D. M. West, 4th Ed., Holt-Saunders, New York, 1982.

2. Many general statistics books have been written. Among them, *Facts from Figures* by M. J. Moroney (Penguin Books) is a long-standing and engagingly written classic. More recently published, and again very readable, is *Statistics in Action* by Peter Sprent (Penguin Books). A further large number of statistics texts are oriented towards scientific (not necessarily chemical) users; *Statistics for Technology* by Christopher Chatfield (Chapman & Hall, London) is an excellent example. Such books provide a quite comprehensive theoretical background to many of the methods described in the present volume.

3. All users of statistics require a good set of statistical tables. *Elementary Statistics Tables* by Henry R. Neave (Allen & Unwin, London) is strongly recommended, in view of its clear explanations of the tables and the associated statistical tests.

4. Several statutory and professional bodies in the U.K., the U.S.A. and elsewhere publish manuals and monographs on standard and recommended statistical procedures and on standard materials. Examples include the British Standards Institute (e.g. BS 5497, *Precision of Test Methods, Part 1 — Guide to the Determination of Repeatability and Reproducibility of a Standard Test Method;* and BS 1797, *Calibration of Volumetric Glassware*): the Association of Official Analytical Chemists (W. J. Youden and E. H. Steiner, *Statistical Manual of the AOAC,* 1975): the International Organisation for Standardisation (ISO) (e.g. IS 4259, *Petroleum Products — Determination and Application of Precision Data in Relation to Methods of Test*): and the American Society for Testing and Materials (ASTM) (e.g. Standard Specifications E617-78, *Laboratory Weights and Precision Mass Standards,* and E694-79, *Volumetric Ware*.

5. Monographs describing standard procedures and materials in specific areas of study are exemplified by A. Knowles and C. Burgess (eds.) *Standards in Absorption Spectrometry,* and J. N. Miller (ed.) *Standards in Fluorescence Spectrometry,* both published by Chapman & Hall in conjunction with the UV Spectrometry Group. The National Bureau of Standards in the U.S.A. publishes monographs to accompany its extensive range of standard materials. A recent example, describing the use of quinine sulphate as a standard in fluorescence spectrometry, is NBS Special Publication No. 260-64 (1980) by R. A. Velapoldi and K. D. Mielenz.

6. Several books describing statistical methods and providing listings of suitable BASIC computer programs have become available recently. These include two books by J. D. Lee and T. D. Lee (*Statistics and Computer Methods in BASIC* and *Statistics and Numerical Methods in BASIC for Biologists,* both published by Van Nostrand Reinhold); *Basic Statistical Computing* by D. Cooke, A. H. Craven and G. M. Clarke, published by Arnold; and *BASIC Microcomputing and Biostatistics* by D. W. Rogers, published by Humana Press Inc.

## EXERCISES

1. A standard sample of pooled human blood serum contains 42.0 g of albumin per litre. Five laboratories (A–E) each do six determinations (on the same day) of the albumin  concentration, with the following results (g/l. throughout):

| | | | | | | |
|---|---|---|---|---|---|---|
| A | 42.5 | 41.6 | 42.1 | 41.9 | 41.1 | 42.2 |
| B | 39.8 | 43.6 | 42.1 | 40.1 | 43.9 | 41.9 |
| C | 43.5 | 42.8 | 43.8 | 43.1 | 42.7 | 43.3 |
| D | 35.0 | 43.0 | 37.1 | 40.5 | 36.8 | 42.2 |
| E | 42.2 | 41.6 | 42.0 | 41.8 | 42.6 | 39.0 |

Comment on the accuracy and precision of each of these sets of results.

2. Using the same sample and method as in question 1, laboratory A makes six further determinations of the albumin concentration, this time on six successive days. The values obtained are 41.5, 40.8, 43.3, 41.9, 42.2 and 41.7 g/l. Comment on these results.

3. The number of binding sites per molecule in a sample of monoclonal antibodies is determined four times, with results of 1.95, 1.95, 1.92 and 1.97. Comment on the accuracy and precision of these results.

4. Discuss the degrees of accuracy and precision desirable in the following analyses:

    (i) Determination of the lactate concentration of human blood samples.
    (ii) Determination of uranium in an ore sample.
    (iii) Determination of a drug in blood plasma after an overdose.
    (iv) Study of the stability of a colorimetric reagent by determination of its absorbance at a single wavelength over a period of several weeks.

5. For each of the following experiments, try to identify the major probable sources of random and systematic errors, and consider how such errors may be minimized:

    (i) The iron content of a large lump of ore is determined by taking a single small sample, dissolving it in acid, and titrating with ceric sulphate after reduction of Fe(III) to Fe(II).
    (ii) The same sampling and dissolution procedure is used as in (i) but the iron is determined colorimetrically after addition of a chelating reagent and extraction of the resulting coloured complex into an organic solvent.
    (iii) The sulphate content of an aqueous solution is determined gravimetrically with barium chloride as the precipitant.

# 2

# Errors in classical analysis – statistics of repeated measurements

## 2.1 MEAN AND STANDARD DEVIATION

In this chapter some fundamental statistical concepts are introduced and applied to the situation common in classical analysis, i.e. repeated measurements of the same quantity are made. In Chapter 1 the various types of error were illustrated by considering the results of five replicate titrations done by each of four students: these results are reproduced below.

| Student | Results (ml) | | | | |
|---------|-------|-------|-------|-------|-------|
| A | 10.08 | 10.11 | 10.09 | 10.10 | 10.12 |
| B | 9.88 | 10.14 | 10.02 | 9.80 | 10.21 |
| C | 10.19 | 9.79 | 9.69 | 10.05 | 9.78 |
| D | 10.04 | 9.98 | 10.02 | 9.97 | 10.04 |

Two criteria were used to compare these results, the average value and the degree of spread. The average value used was the **arithmetic mean, $\bar{x}$,** (usually abbreviated to the **mean**) which is the sum of all the measurements divided by the number of measurements:

$$\bar{x} = \sum_i x_i/n \qquad (2.1)$$

The most useful measure of spread is the **standard deviation ($s$)**. This is defined by the formula

$$s = \sqrt{\sum_i (x_i - \bar{x})^2/(n-1)} \qquad (2.2)$$

The calculation of the standard deviation of A's results is shown in Table 2.1.

**Table 2.1** – Calculation of mean and standard deviation for student A's results

| | $x_i$(ml) | $(x_i - \overline{x})^2$ |
|---|---|---|
| | 10.08 | 0.0004 |
| | 10.11 | 0.0001 |
| | 10.09 | 0.0001 |
| | 10.10 | 0.0000 |
| | 10.12 | 0.0004 |
| Total | 50.50 | 0.0010 |

$$\overline{x} = 50.50/5 = 10.10$$
$$s = \sqrt{0.0010/4} = 0.016 \text{ ml}$$

The reader should check that the standard deviations of the results of students B, C and D are 0.17, 0.21 and 0.029 ml respectively, giving quantitative confirmation of the assessments of precision made in Chapter 1.

Many pocket calculators will give the results of these calculations if the values of $x$ are keyed in. However, care must be taken that the right key is pressed to obtain the standard deviation. Some calculators give two different values for the standard deviation, one calculated by using Eq. (2.2) and the other calculated with the denominator of this equation, i.e. $(n-1)$, replaced by $n$. (The reason for these two different forms is explained below, p. 36.) Obviously, for large values of $n$ the difference is negligible.

Unfortunately, in calculating a standard deviation, the calculator may round off numbers so that an erroneous value may be obtained for the standard deviation (even zero). This usually happens when the difference between the input values occurs at the fourth or a subsequent significant digit, depending on the make of calculator. For example, many calculators give the standard deviation of the three values 100.000, 100.001 and 100.002 as zero, whereas it is in fact 0.000816 (to 3 significant figures). This problem can be overcome by **coding** the values, i.e. subtracting a fixed amount from each, to give, in this case, 0.000, 0.001 and 0.002. Since the standard deviation measures the spread about the mean, the standard deviation of these coded values is the same as that of the original values. (The mean of the original values is given by adding 100.000 to the mean of the coded values.) Coding also reduces the labour and possible mistakes attendant on punching information into a calculator. In this particular example, the calculation could have been made even quicker and less prone to error by taking the coded values as 0, 1 and 2: in such cases especial care must be taken in uncoding the output of the calculator.

The calculation in Table 2.1 was very simple because the values of $(x_i - \overline{x})^2$ could be worked out mentally. This is not usually so and an alternative form of Eq. (2.2) can be used to simplify the arithmetic if a preprogrammed calculator is not available:

$$s = \sqrt{\frac{\sum_i x_i^2}{(n-1)} - \frac{(\sum_i x_i)^2}{n(n-1)}} \qquad (2.3)$$

The mean and the standard deviation can also be calculated on a computer with a BASIC program of only a few lines. Such a program would only be of value if the computer itself was collecting the data or a large amount of data was involved. The writing of such a program is left to the reader — example programs are given in the books in the bibliography at the end of chapter 1.

The square of $s$ is a very important statistical quantity known as the **variance**; its value will become apparent later in the chapter when we consider the propagation of errors. Also widely used is the **coefficient of variation (CV)**, also known as the **relative standard deviation (RSD)** and given by $100s/\bar{x}$. The CV or RSD, the units of which are obviously per cent, is an example of a **relative error**, i.e. an error estimate divided by an estimate of the absolute value of the measured quantity. Relative errors are frequently used in the comparison of the precisions of results which have different units or magnitudes, and are again important in calculations of error propagation.

## 2.2  DISTRIBUTION OF ERRORS

Although the standard deviation gives a measure of the spread of a set of results about the mean value, it does not indicate the way in which the results are distributed. To illustrate this we need a large number of measurements such as those in Table 2.2. This gives the results of 50 determinations of the nitrate ion concentration in a particular water specimen, given to two significant figures.

**Table 2.2** — Results of 50 determinations of nitrate ion concentration, in $\mu$g/ml

| | | | | | | | | | |
|---|---|---|---|---|---|---|---|---|---|
| 0.51 | 0.51 | 0.51 | 0.50 | 0.51 | 0.49 | 0.52 | 0.53 | 0.50 | 0.47 |
| 0.51 | 0.52 | 0.53 | 0.48 | 0.49 | 0.50 | 0.52 | 0.49 | 0.49 | 0.50 |
| 0.49 | 0.48 | 0.46 | 0.49 | 0.49 | 0.48 | 0.49 | 0.49 | 0.51 | 0.47 |
| 0.51 | 0.51 | 0.51 | 0.48 | 0.50 | 0.47 | 0.50 | 0.51 | 0.49 | 0.48 |
| 0.51 | 0.50 | 0.50 | 0.53 | 0.52 | 0.52 | 0.50 | 0.50 | 0.51 | 0.51 |

These can be summarized in a **frequency table** (Table 2.3)

**Table 2.3** — Frequency table for measurements of nitrate ion concentration

| Nitrate ion concentration ($\mu$g/ml) | Frequency |
|---|---|
| 0.46 | 1 |
| 0.47 | 3 |
| 0.48 | 5 |
| 0.49 | 10 |
| 0.50 | 10 |
| 0.51 | 13 |
| 0.52 | 5 |
| 0.53 | 3 |

This table thus indicates that, in Table 2.2, the value 0.46 μg/ml appears once, the value 0.47 μg/ml appears three times, and so on. The reader can check that the mean of these results is 0.500 μg/ml and the standard deviation is 0.0165 μg/ml. (These values are arbitrarily given to three significant figures: further discussion of this important aspect of the presentation of results appears later in the chapter). The distribution of the results can be most easily appreciated by drawing a **histogram** as in Fig. 2.1. This shows that the distribution of the measurements is roughly symmetrical about the mean, with the measurements clustered towards the centre.

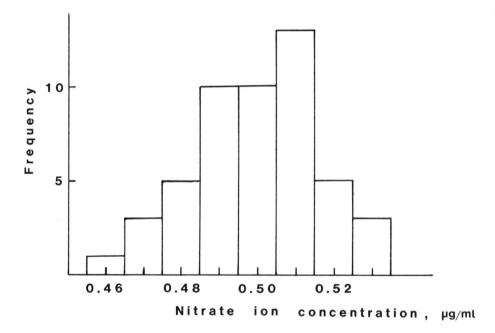

Fig. 2.1 – Histogram of the nitrate ion concentration data in Table 2.3.

This set of 50 measurements is a **sample** from the very large (in theory infinite) number of measurements which we could make of the nitrate ion concentration. The set of possible measurements is called the **population**. *If there are no systematic errors,* then the mean of this population, denoted by $\mu$, is the true value of the nitrate ion concentration which we are trying to determine. The mean, $\bar{x}$, of the sample gives us an estimate of $\mu$. Similarly, the population has a standard deviation, denoted by $\sigma$. The value of the standard deviation, $s$, of the sample gives us an estimate of $\sigma$. Use of Eq. (2.2) gives an unbiased estimate of $\sigma$, i.e. as the sample size tends to infinity $s$ tends to $\sigma$. If $n$, rather than $n - 1$, is used in the denominator of the equation, the value of $s$ obtained is an under-estimate of $\sigma$.

The measurements of nitrate ion concentration given in Table 2.2 have only certain discrete values because of the limitations of the method of measurement. In theory a concentration could take any value, so to describe the form of the population from which the sample was taken a continuous curve is needed. The mathematical model usually used is the **normal** or **Gaussian distribution** which is described by the equation

$$y = \frac{\exp\{-(x-\mu)^2/2\sigma^2\}}{\sigma\sqrt{2\pi}} \tag{2.4}$$

and its shape is shown in Fig. 2.2. There is no need to remember this complicated formula but some of its general properties are important. The curve is symmetrical about $\mu$ and the greater the value of $\sigma$ the greater the spread of the curve, as shown in Fig. 2.3. More detailed analysis shows that, whatever the values of $\mu$

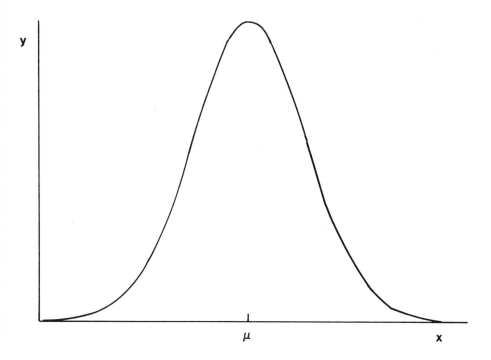

Fig. 2.2 – The normal distribution, $y = \exp[-(x-\mu)^2/2\sigma^2]/\sigma\sqrt{2\pi}$. The mean is indicated by $\mu$.

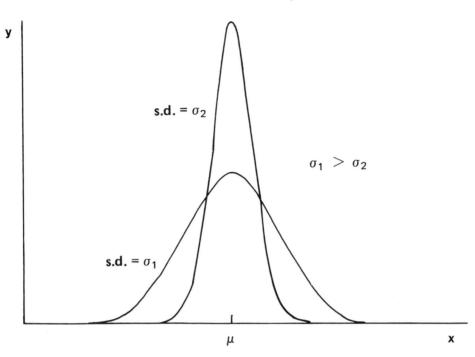

Fig. 2.3 – Normal distributions with the same mean but different values of the standard deviation.

and $\sigma$, approximately 68% of the population values lie within $\pm 1\sigma$ of the mean, approximately 95% of the values lie within $\pm 2\sigma$ of the mean and approximately 99.7% of the values lie within $\pm 3\sigma$ of the mean (Fig. 2.4). This would mean that, if the nitrate concentrations given in Table 2.3 are normally distributed, about 68% should lie in the range 0.483–0.517, about 95% in the range 0.467–0.533 and 99.7% in the range 0.450–0.550. In fact 33 of the 50 results (66%) lie between 0.483 and 0.517, 49 (98%) between 0.467 and 0.533, and all the results lie between 0.450 and 0.550, so the agreement with theory is fairly good.

Although it cannot be proved that repeated measurements of a single analytical quantity are always normally distributed, there is considerable evidence that this assumption is generally at least approximately true. Moreover we shall see, when we come to look at sample means, that any departure of a population from normality is not usually important in the context of the statistical tests most frequently used.

The normal distribution is not only applicable where repeated measurements are made on the same specimen. It also often fits the distribution of results obtained when the same quantity is measured for different samples. For example if we measured the concentrations of albumin in blood sera taken from healthy

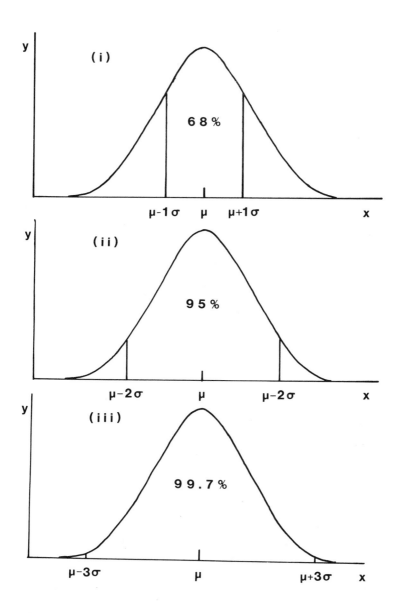

Fig. 2.4 — Properties of the normal distribution: (i) approximately 68% of values lie within ± 1σ of the mean; (ii) approximately 95% of values lie within ±2σ of the mean; (iii) approximately 99.7% of values lie within ±3σ of the mean.

adult humans we would find that the results were approximately normally distributed. However, in this second type of population, i.e. one measurement on each of a number of specimens, other distributions are not uncommon. In particular the so-called log-normal distribution is frequently encountered: in this distribution the *logarithms* of the concentrations (or other characteristics), when plotted against frequency, give a normal distribution curve. For example the antibody concentration in human blood sera is approximately log-normally distributed (Fig. 2.5), and the particle sizes of the droplets formed by the nebulizers used in flame spectrometry may also follow this distribution.

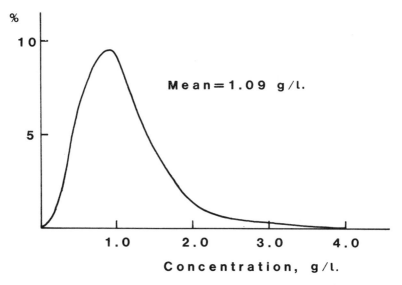

Fig. 2.5 — An approximately log–normal distribution: concentrations of serum immunoglobulin M antibody in male subjects. Note the asymmetry of the curve.

## 2.3 THE SAMPLING DISTRIBUTION OF THE MEAN

We have already seen that the mean of a sample of measurements provides us with an estimate of the true value, $\mu$, of the quantity we are trying to measure. Since, however, the individual measurements are distributed about the true value, with a spread which depends on the precision, it is most unlikely that the mean of the sample is *exactly* equal to the true value. For this reason it is more useful to give a range of values within which we are almost certain the true value lies. The width of this range depends on two factors. The first is the precision of the individual measurements, which in turn depends on the variance of the population. The second is the number of measurements in the sample. The very fact that we repeat measurements implies we have more confidence in the mean

of several values than in a single one. Most people will feel that the more measurements we make, the more reliable our estimate of $\mu$, the true value, will be. To
pursue this idea, let us return to the nitrate ion determination described in
Section 2.2. In practice it would be most unusual to make 50 repeated measurements in such a case. A more likely number would be 5 and we can see how
the means of samples of this size are spread about $\mu$ by treating the results in
Table 2.2 as ten samples each containing 5 results. Taking each column as one
sample, the means are 0.506, 0.504, 0.502, 0.496, 0.502, 0.492, 0.506, 0.504,
0.500, 0.486. It is at once obvious that these means are more closely clustered
than the original measurements. Just as the original measurements were a sample
from an infinite population of possible measurements, so these means are a
sample from the possible means of samples of 5 measurements from the whole
population. The distribution of these sample means is called the **sampling
distribution of the mean**. Its mean is the same as the mean of the original population. Its standard deviation is called the **standard error of the mean (s.e.m.)**.
There is an exact mathematical relationship between it and the standard deviation,
$\sigma$, of the distribution of the individual measurements, which is independent of
the way in which they are distributed. If $n$ is the sample size this relationship is:

$$\text{s.e.m.} = \sigma/\sqrt{n} \tag{2.4}$$

As we expect intuitively, the larger $n$ is, the smaller the spread of the sample
means about $\mu$.

Another property of the sampling distribution of the mean is that, *even if
the original population is not normally distributed,* the sampling distribution of
the mean tends to the normal distribution as $n$ increases. This result is known as
the **central limit theorem**. This theorem is of the utmost importance because
many statistical tests are performed on the mean and assume that it is normally
distributed. Since in practice we can assume that distributions of repeated
measurements are at least approximately normally distributed, it is reasonable
to assume that the means of quite small samples (say $> 5$) are normally distributed.

## 2.4 CONFIDENCE LIMITS OF THE MEAN

Now that we know the form of the sampling distribution of the mean we can
return to the problem of using a sample to define a range within which we may
reasonably assume the true value lies. (Remember that in doing this we are
assuming systematic errors to be absent.) Such a range is known as a **confidence
interval** and the extreme values of the range are called the **confidence limits**.
The term 'confidence' implies that we can assert with a given degree of
confidence, i.e. a certain probability, that the confidence interval *does* include
the true value. The size of the confidence interval will obviously depend on how
certain we want to be that it includes the true value: the greater the certainty,
the *greater* the interval required.

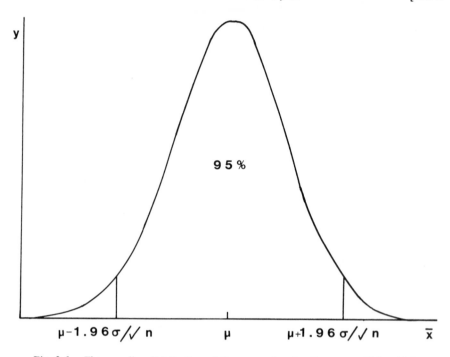

Fig. 2.6 – The sampling distribution of the mean, showing the range within which 95% of sample means lie.

Figure 2.6 shows the sampling distribution of the mean for samples of size $n$ results. Assuming from now on that the distribution is normal, this shows that 95% of the sample means will lie in the range given by:

$$\mu - 1.96(\sigma/\sqrt{n}) < \bar{x} < \mu + 1.96(\sigma/\sqrt{n}) \tag{2.5}$$

(The exact value 1.96 is used in this equation rather than the approximate value, 2, which is often used.)

In practice, however, we usually have one sample, of known mean, and we require a range for $\mu$, the true value. Equation (2.5) can be rearranged to give this:

$$\bar{x} - 1.96(\sigma/\sqrt{n}) < \mu < \bar{x} + 1.96(\sigma/\sqrt{n}) \tag{2.6}$$

Equation (2.6) gives the **95% confidence interval** of the mean. Similarly the 99.7% confidence interval is given by:

$$\bar{x} - 2.97(\sigma/\sqrt{n}) < \mu < \bar{x} + 2.97(\sigma/\sqrt{n}) \tag{2.7}$$

Another confidence interval commonly used is the 99% confidence interval which is given by:

$$\bar{x} - 2.58(\sigma/\sqrt{n}) < \mu < \bar{x} + 2.58(\sigma/\sqrt{n}) \tag{2.8}$$

Equation (2.6) can be used to calculate the 95% confidence limits for the nitrate ion concentration. We have $\bar{x} = 0.500$ and $n = 50$. The only quantity in the equation which we do not know is $\sigma$. For *larger* samples such as this one, $s$ gives a sufficiently accurate estimate of $\sigma$ to replace it. Thus the 95% confidence interval for the nitrate ion concentration is:

$$0.500 - 1.96 \times 0.0165/\sqrt{50} < \mu < 0.500 + 1.96 \times 0.0165/\sqrt{50}$$

giving the confidence limits as:

$$\mu = 0.500 \pm 0.0046 \ \mu g/ml$$

As the sample size gets smaller, the uncertainty introduced by using $s$ to estimate $\sigma$ increases. To allow for this the equation used to calculate the confidence limits is modified to:

$$\mu = \bar{x} \pm t\,(s/\sqrt{n}) \tag{2.9}$$

The appropriate value of $t$ depends both on $(n - 1)$, which is known as the number of **degrees of freedom** (usually given the symbol $\nu$) and the degree of confidence required. [The term degrees of freedom refers to the number of *independent* deviations $(x_i - \bar{x})$ which are used in calculating $s$. In this case the number is $(n - 1)$, because when $(n - 1)$ deviations are known, the last can be deduced by using the obvious result $\sum_i (x_i - \bar{x}) = 0$]. Values of $t$ are given in Table 2.4.

**Table 2.4** – Values of $t$ for confidence intervals

| Degrees of freedom | Values of $t$ for confidence interval of | |
|---|---|---|
| | 95% | 99% |
| 1 | 12.71 | 63.66 |
| 2 | 4.30 | 9.92 |
| 3 | 3.18 | 5.84 |
| 4 | 2.78 | 4.60 |
| 5 | 2.57 | 4.03 |
| 10 | 2.23 | 3.17 |
| 20 | 2.09 | 2.85 |
| 30 | 2.04 | 2.75 |
| 50 | 2.01 | 2.68 |
| 100 | 1.98 | 2.63 |

It can be seen from this table that for sample sizes greater than 50 the values of $t$ are very close to the values 1.96 and 2.58 used in Eqs. (2.6) and (2.8) respectively. This confirms the validity of the assumption used above in calculating the confidence limits for the nitrate concentration. The use of this table can be illustrated by an example.

*Example.* The sodium ion content of a urine sample was determined by use of an ion-selective electrode. The following values were obtained: 102, 97, 99, 98, 101, 106mM. What are the 95% and 99% confidence limits for the sodium ion concentration?

The mean and standard deviation of these values are 100.5mM and 3.27mM respectively. There are 6 measurements and therefore 5 degrees of freedom. From Table 2.4 the value of $t$ for calculating the 95% confidence limits is 2.57 and from Eq. (2.9) the 95% confidence limits are

$$\mu = 100.5 \pm 2.57 \times 3.27/\sqrt{6}$$

$$= 100.5 \pm 3.4\text{m}M$$

Similarly the 99% confidence limits are

$$\mu = 100.5 \pm 4.03 \times 3.27/\sqrt{6}$$

$$= 100.5 \pm 5.4\text{m}M$$

If the sample comes from a population which is known to be log normally distributed then the individual values should be transformed by taking their logarithms before the confidence limits are calculated. Since the transformed values will be normally distributed, the confidence limits of the mean can then be calculated as described above. The antilogarithms of the calculated confidence limits give the confidence limits appropriate to the original data. Note, however, that the values of the mean and standard deviation of the sample are calculated from the original untransformed values. As a result of the asymmetry of the log normal distribution (see Fig. 2.5) the confidence interval of the mean, calculated as described above, is not symmetrical about the mean.

## 2.5 PRESENTATION OF RESULTS

As has already been emphasized, no quantitative experimental result is of any value unless it is accompanied by an estimate of the errors involved in its measurement. The usual practice is to quote the mean as the estimate of the quantity measured and the standard deviation as the estimate of the precision. Less commonly, the standard error of the mean is sometimes quoted instead of the standard deviation, or the result is given in the form of the 95% confidence limits of the mean. In all cases it is obviously important to state the form used and, *provided that the value of n is given*, the three forms can be easily inter-converted by using Eqs. (2.4) and (2.9).

A related aspect of presenting results is the rounding-off of the answer. The important principle here is that the *number of significant figures given indicates the precision of the experiment.* It would clearly be absurd, for example, to give the result of a titrimetric analysis as 0.107846M — no analyst could achieve the implied precision of 0.000001 in ca. 0.1, i.e. 0.001%. In practice it is usual to quote all the certain significant figures, plus the first uncertain figure. For example, the mean of the values 10.09, 10.11, 10.09, 10.10 and 10.12 is 10.102, and their standard deviation is 0.0130. Clearly there is uncertainty in the second decimal place; the results are all 10.1 to one decimal place, but disagree in the second decimal place. By the suggested method the result would be quoted as:

$$\bar{x} \pm s = 10.10 \pm 0.01 \ (n = 5)$$

If it was felt that this resulted in an unacceptable rounding-off of the standard deviation, then the result could be given as:

$$\bar{x} \pm s = 10.10_2 \pm 0.01_3 \ (n = 5)$$

where the use of a subscript indicates that the digit is given only to avoid loss of information. The reader could decide whether it was useful or not.

Similarly, when confidence limits are calculated [see Eq. (2.9)], there is no point in giving the value of $ts/\sqrt{n}$ to more than 2 significant figures. The value of $\bar{x}$ should then be given to the corresponding number of decimal places.

The number of significant figures quoted is sometimes used instead of a specific estimate, to indicate the precision of a result. For example $0.1046M$ is taken to mean that the figures in the first three decimal places are certain but there is doubt about the fourth. However, since the uncertainty in the last figure could be anything from 0.00005 to 0.0005, this method gives a poor estimate of precision and is not to be recommended. Sometimes the uncertainty in the last figure is emphasized by using the formats $0.104(6)M$ or, more usually, $0.104_6M$, but it remains preferable to give a specific estimate of precision such as the standard deviation.

One problem which arises is whether a 5 should be rounded up or down. For example, if 9.65 is rounded to one decimal place, should it become 9.6 or 9.7? It is evident that the results will be biased if a 5 is always rounded up; this bias can be avoided by rounding the 5 to the nearest *even* number, giving, in this case, 9.6. Analogously, 4.75 is rounded to 4.8.

When several measured quantities are to be used to calculate a final result (see Section 2.7) these quantities should not be rounded-off too much or a needless loss of precision will result. A good rule is to keep one digit beyond the last significant figure and leave further rounding until the final result is reached. The same advice applies when the mean and standard deviation are used to apply a statistical test such as the $F$- and $t$-tests (see Chapter 3): the unrounded values of $x$ and $s$ should be used in the calculations.

## 2.6 OTHER USES OF CONFIDENCE LIMITS

Confidence intervals can be used as a test for systematic errors as shown in the following example.

*Example*. The absorbance scale of a spectrometer is tested at a particular wavelength by using a standard solution which has an absorbance given as 0.470. Ten measurements of the absorbance with the spectrometer gave $\bar{x} = 0.461$ and $s = 0.003$. Find the 95% confidence interval for the mean absorbance as measured by the spectrometer, and hence decide whether a systematic error is present.

The 95% confidence limits for the absorbance as measured by the spectrometer are [Eq. (2.9)] :

$$\mu = \bar{x} \pm t(s/\sqrt{n})$$
$$= 0.461 \pm 2.26 \times 0.003/\sqrt{10}$$
$$= 0.461 \pm 0.002$$

(The value of $t$ was obtained from Table A.1 in the appendices, which is a more complete version of Table 2.4).

Since this confidence interval does not include the known absorbance of 0.470, it is likely that there is a systematic error.

Another approach to the same problem is introduced in the next chapter.

Confidence limits can also be used when measurements are made on each of a number of specimens. Suppose for example that the mean weight of a tablet in a very large batch is required: it would be too time-consuming to weigh each tablet. Similarly, if a destructive method of analysis such as atomic-absorption spectrometry is to be used to analyse the batch to assess a mean content, it is clearly impossible to examine every tablet. In both cases, a sample could be taken from the batch (which is such instances forms the population) and from the mean and standard deviation of the sample, a confidence interval could be found for the mean value of the quantity measured.

## 2.7 PROPAGATION OF RANDOM ERRORS

In experimental work, the quantity to be determined is often calculated from a combination of observable quantities. We have already seen, for example, that even a relatively simple operation such as a titrimetric analysis involves several stages, each of which will be subject to errors (see Chapter 1). The final calculation may involve taking the sum, difference, product or quotient of two or more quantities, or the raising of any quantity to a power.

It is most important to note that the procedures used for combining random and systematic errors are completely different. This is because random errors to some extent cancel each other out, whereas every systematic error occurs in a definite and known sense. Suppose for example that the final result of an experiment, $x$, is given by $x = a + b$. If $a$ and $b$ each have a systematic error of $+1$, it is clear that the systematic error in $x$ is $+2$. If, however, $a$ and $b$ each have a random error of $\pm 1$, the random error in $x$ is *not* $\pm 2$: this is because there will be occasions when the random error in $a$ is positive while that in $b$ is negative (or vice versa).

This section deals only with the propagation of random errors (systematic errors are considered below). If the precision of each observation is known, then simple mathematical rules can be used to estimate the precision of the final result. These rules can be summarized as follows.

### (i) *Linear combinations*
In this case the final value, $y$, is calculated from a linear combination of measured quantities $a$, $b$, $c$, etc. by:

$$y = k + k_a a + k_b b + k_c c + \ldots \tag{2.10}$$

where $k$, $k_a$, $k_b$, $k_c$, etc. are constants. Variance (defined as the square of the standard deviation) has the important property that the variance of a sum or difference of *independent* quantities is equal to the sum of their variances. It can be shown that if $\sigma_a$, $\sigma_b$, $\sigma_c$, etc. are the standard deviations of $a$, $b$, $c$, etc. the standard deviation of $y$, $\sigma_y$, is given by:

$$\sigma_y = \sqrt{(k_a \sigma_a)^2 + (k_b \sigma_b)^2 + (k_c \sigma_c)^2 + \ldots} \tag{2.11}$$

*Example.* In a titration the initial reading on the burette is 3.51 ml and the final reading is 15.67 ml, both with a standard deviation of 0.02 ml. What is the volume of titrant used and what is its standard deviation?

$$\text{Volume used} = 15.67 - 3.51 = 12.16 \text{ ml}$$

$$\text{Standard deviation} = \sqrt{(0.02)^2 + (0.02)^2} = 0.028 \text{ ml.}$$

This example illustrates the important point that the standard deviation for the final result is larger than the standard deviations of the individual burette readings even though the volume used is calculated from a difference, but is less than the sum of the standard deviations.

### (ii) *Multiplicative expressions*
If $y$ is calculated from an expression of the type:

$$y = kab/cd \tag{2.12}$$

(where $a$, $b$, $c$ and $d$ are independent measured quantities and $k$ is a constant) then there is a relationship between the squares of the *relative* standard deviations:

$$\frac{\sigma_y}{y} = \sqrt{\left(\frac{\sigma_a}{a}\right)^2 + \left(\frac{\sigma_b}{b}\right)^2 + \left(\frac{\sigma_c}{c}\right)^2 + \left(\frac{\sigma_d}{d}\right)^2} \tag{2.13}$$

*Example.* The quantum yield of fluorescence, $\phi$, is calculated from the expression:

$$\phi = I_f/kcII_0\epsilon$$

where the quantities involved are defined below, with an estimate of their relative standard deviations in brackets:

$I_0$ = incident light intensity (0.5%)
$I_f$ = fluorescence light intensity (2%)
$\epsilon$  = molar absorptivity (1%)
$c$  = concentration (0.2%)
$l$  = path-length (0.2%)
$k$   is an instrument constant.

From Eq. (2.13), the relative standard deviation of $\phi$ is given by:

$$\text{r.s.d.} = \sqrt{2^2 + 0.2^2 + 0.2^2 + 0.5^2 + 1^2}$$

$$= \sqrt{4 + 0.04 + 0.04 + 0.25 + 1}$$

$$= \sqrt{5.33}$$

$$= 2.3\%$$

It can be seen that the relative standard deviation in the final result is not much larger than the largest relative standard deviation used to calculate it (i.e. 2% for $I_f$). This is mainly a consequence of the squaring of the relative standard deviations and illustrates an important general point: any efforts to improve the precision of an experiment need to be directed towards improving the precision of the least precise values. As a corollary to this, there is no point in wasting effort in increasing the precision of the most precise values. This is not to say that small errors are unimportant: small errors at many stages of an experiment, such as the titrimetric analysis discussed in detail in Chapter 1, will produce an appreciable error in the final result.

It is important to note that when a quantity is raised to a power, e.g. $b^3$, then the error is not calculated as for a multiplication, i.e. $b \times b \times b$, because the quantities involved are not independent. If the relationship is:

$$y = b^n \tag{2.14}$$

then the standard deviations of $y$ and $b$ are related by:

$$\frac{\sigma_y}{y} = \left| \frac{n \, \sigma_b}{b} \right| \tag{2.15}$$

(The modulus sign | | means that the magnitude of the enclosed quantity is taken without respect to sign, e.g. $|-2| = 2$).

(iii) *Other functions*
If $y$ is a general function of $x$:

$$y = f(x) \tag{2.16}$$

then the standard deviations of $x$ and $y$ are related by:

$$\sigma_y = \left| \sigma_x \cdot \frac{dy}{dx} \right| \tag{2.17}$$

*Example.* The absorbance, $A$, of a solution is given by:

$$A = -\log(T)$$

where $T$ is the transmittance. If the measured value of $T$ is 0.501 with a standard deviation of 0.001, calculate $A$ and its standard deviation.

We have:

$$A = -\log 0.501 = 0.300$$

Also:

$$dA/dT = -\log e/T$$
$$= -0.434/T$$

so that from Eq. (2.17):

$$\text{standard deviation of } A = |0.001 \times (-0.434/0.501)| = 0.0008_7$$

## 2.8 PROPAGATION OF SYSTEMATIC ERRORS

The rules for the combination of systematic errors can also be divided into three groups.

(i) *Linear combinations*
If $y$ is calculated from the measured quantities by using Eq. (2.10), and the systematic errors in $a$, $b$, $c$, etc. are $\Delta a$, $\Delta b$, $\Delta c$, etc. then the systematic error in $y$, $\Delta y$, is calculated from:

$$\Delta y = k_a \Delta a + k_b \Delta b + k_c \Delta c + \dots \tag{2.18}$$

Remember that the systematic errors are either positive or negative and that these signs *must* be included in the calculation of $\Delta y$.

The total systematic error can sometimes be zero. Suppose, for example, a balance with a systematic error of $-0.01$ g is used for the weighings involved in making a standard solution. Since the weight of the solute used is found from the difference between two weighings the systematic errors cancel out. Carefully considered procedures, such as this, can often minimize the systematic errors, as described in Chapter 1.

### (ii) *Multiplicative expressions*

If $y$ is calculated from the measured quantities by using Eq. (2.12) then *relative* systematic errors are used:

$$(\Delta y/y) = (\Delta a/a) + (\Delta b/b) + (\Delta c/c) + (\Delta d/d) \qquad (2.19)$$

When a quantity is raised to some power, then Eq. (2.15) is used with the modulus sign omitted and the standard deviations replaced by systematic errors.

### (iii) *Other functions*

The corresponding equation is identical to Eq. (2.17) but with the modulus sign omitted and the standard deviations replaced by systematic errors.

### BIBLIOGRAPHY

O. L. Davies and P. L. Goldsmith, *Statistical Methods in Research and Production,* Longmans, London, 1982. Gives a more detailed treatment of the subject matter of this chapter.
D. A. Skoog and D. M. West, *Fundamentals of Analytical Chemistry,* 4th Ed., Holt Saunders, New York, 1982. Describes the use of statistics in evaluating analytical data.
J. Topping, *Errors of Observation and their Treatment,* Chapman & Hall, London, 1962. Gives a fuller treatment of the theory of errors and discusses the theory that errors are normally distributed.

### EXERCISES

1. To investigate the reproducibility of a method for the determination of selenium in foods, nine measurements were made on a single batch of brown rice, with the following results:

| Sample | Selenium, $\mu g/g$ |
|--------|---------------------|
| 1 | 0.07 |
| 2 | 0.07 |
| 3 | 0.08 |
| 4 | 0.07 |
| 5 | 0.07 |
| 6 | 0.08 |
| 7 | 0.08 |
| 8 | 0.09 |
| 9 | 0.08 |

(T. Moreno-Domínguez, C. García-Moreno and A. Mariné-Font, *Analyst,* 1983, **108,** 505).
    Calculate the mean, standard deviation and relative standard deviation of these results.

2. Seven measurements of the pH of a buffer solution gave the following results:

$$5.12, \ 5.20, \ 5.15, \ 5.17, \ 5.16, \ 5.19, \ 5.15.$$

    Calculate (i) the 95% and (ii) the 99% confidence limits for the true pH. (Assume that there are no systematic errors.)

3. Ten analyses of the concentration of albumin gave a mean of 20.9 g/l. and a standard deviation of 0.45 g/l. (J. W. Foote and H. T. Delves, *Analyst*, 1983, **108**, 492).

Calculate the 95% confidence limits of the mean.

4. The concentration of lead in the blood stream was measured for a sample of 50 children from a large school near a busy main road. The sample mean was 10.1 ng/ml and the standard deviation was 0.6 ng/ml. Calculate the 95% confidence interval for the mean lead concentration for all the children in the school.

About how big should the sample have been to reduce the range of the confidence interval to 0.2 ng/ml (i.e. ±0.1 ng/ml)?

5. In an investigation into the accuracy and precision of a method for determining arprinocid in feed premixes, six replicate determinations were made on a premix formulated to contain 10.2% arprinocid. The results were:

$$10.4, \ 10.4, \ 10.6, \ 10.3, \ 10.5, \ 10.5\%.$$

(J. D. Stong and D. W. Fink, *Analyst*, 1982, **107**, 113).

Calculate the mean, standard deviation and 95% and 99% confidence limits of the mean. Is the formulated value of 10.2% within (i) the 95%, (ii) the 99% confidence limits?

6. Measurement of the haptoglobin concentration in blood serum taken from a random sample of eight healthy adults gave the following values:

$$1.82, \ 3.32, \ 107, \ 1.27, \ 0.49, \ 3.79, \ 0.15, \ 1.98 \ g/l.$$

Calculate the mean and standard deviation of these results.

Assuming the haptoglobin is log-normally distributed in the population as a whole, calculate the 95% confidence interval for the mean haptoglobin concentration for the whole population.

7. Ten measurements of the ratio of two peak areas in a liquid chromatography experiment gave the following values:

$$0.2911, \ 0.2898, \ 0.2923, \ 0.3019, \ 0.2997, \ 0.2961, \ 0.2947, \ 0.2986, \ 0.2902, \ 0.2882.$$

(P. Jonvel and G. Andermann, *Analyst*, 1983, **108**, 411).

Calculate the mean, standard deviation and 99% confidence limits of the mean.

8. A 0.1*M* solution of acid was used to titrate 10 ml of 0.1*M* solution of alkali and the following volumes of acid were recorded.

$$9.88, \ 10.18, \ 10.23, \ 10.39, \ 10.25 \ ml$$

Calculate the 95% confidence limits of the mean and use them to decide if there is any evidence of systematic error.

9. This problem considers the random errors involved in making up a standard solution. A volume of 250 ml of 0.05*M* solution of a reagent of formula weight (relative molecular mass) 40 was made up, the weighing being done by difference. The standard deviation of each weighing was 0.0001 g: what were the standard deviation and relative standard deviation of the weight of reagent used? The standard deviation of the volume of solvent used was 0.05 ml. Express this as a relative standard deviation. Hence calculate the relative standard deviation of the molarity of the solution.

Repeat the calculation for a reagent of formula weight 392.

10. The solubility product of barium sulphate is $1.3 \times 10^{-10}$, with a standard deviation of $0.1 \times 10^{-10}$. Calculate the standard deviation of the calculated solubility of barium sulphate in water.

# 3

# Significance tests

## 3.1 INTRODUCTION

One of the most important properties of an analytical method is that it should be free from systematic error, i.e. the value which it gives for the amount of the analyte should be the *true* value. This property may be tested by applying the method to a standard sample containing a known amount of analyte (Chapter 1). However, as we saw in the last chapter, random errors make it most unlikely that the measured amount would *exactly* equal the known amount even if there were no systematic error. In order to decide whether the difference between the measured and known amounts can be accounted for by these random errors a statistical test known as a **significance test** can be employed. As its name implies, this approach tests whether the difference between the two results is significant, or can be accounted for merely by random variations. Significance tests are widely used in the evaluation of experimental results. This chapter considers several tests which are particularly useful to analytical chemists.

## 3.2 COMPARISON OF AN EXPERIMENTAL MEAN WITH A KNOWN VALUE

In making a significance test we are testing the truth of a hypothesis of the type which is known as a **null hypothesis**. For the example in the previous paragraph we adopt the null hypothesis that the analytical method is *not* subject to systematic error. The term *null* is used to imply that there is *no* difference between the observed and known values other than that which can be attributed to random variation. Assuming that this null hypothesis is true, statistical theory can be used to calculate the probability (i.e. the chance) that the observed difference between the sample mean, $\bar{x}$, and the true value, $\mu$, arises solely as a result of random errors. The lower the probability that the observed difference occurs by chance, the less likely it is that the null hypothesis is true. Usually the null hypothesis is rejected if the probability of the observed difference occurring by chance is *less* than 1 in 20 (i.e. 0.05) and in such a case the difference is said to be **significant at the 0.05 (or 5%) level.** Using this level of significance, we shall, on average, reject the null hypothesis *when it is in fact true* once in 20 times. In order to to be more certain that we make the correct decision a higher level of significance can be used, usually 0.01 or 0.001 (1% or 0.1%). The significance level is indicated by writing, for example, $P$ (i.e. probability) = 0.05. It is important to appreciate that a decision to retain the null hypothesis does not mean it has been *proved* to be true, but only that it has not been demonstrated to be false.

In order to decide whether the difference between $\mu$ and $\bar{x}$ is significant, Eq. (2.9):

$$\mu = \bar{x} \pm (ts/\sqrt{n})$$

(where $n$ is the sample size) is rewritten:

$$t = (\bar{x} - \mu)\sqrt{n}/s \tag{3.1}$$

and a value of $t$ is calculated by substituting the experimental results in this equation. If $|t|$ (i.e. the value of $t$ without regard to sign) *exceeds* a certain **critical value** then the null hypothesis is rejected. The critical value of $|t|$ for a particular significance level is found from Table A.1. For example, for a sample size of 10 (i.e. 9 degrees of freedom) and a significance level of 0.01, the critical value of $|t|$ is 3.25.

*Example.* In a method for determining mercury by cold-vapour atomic absorption the following values were obtained for a standard reference material containing 38.9% mercury:

$$38.9, \ 37.4, \ 37.1\%$$

(P.-K. Hon, O.-W. Lau and M.-C. Wong, *Analyst,* 1983, **108**, 64).
Is there any evidence of systematic error?
The mean of these values is 37.8% and the standard deviation is 0.964%. Adopting the null hypothesis that there is no systematic error, and using Eq. (3.1) gives

$$|t| = |(37.8 - 38.9) \times \sqrt{3}/0.964| = 1.98$$

From Table A.1, for 2 degrees of freedom the critical value of $|t|$ is 4.3 ($P = 0.05$). Since the observed value of $|t|$ is less than the critical value the null hypothesis is retained: there is no evidence of systematic error. Note again that this does not mean that there *are* no systematic errors, only that they have not been demonstrated.

## 3.3 COMPARISON OF THE MEANS OF TWO SAMPLES

Another way in which the results of a new analytical method may be tested is by comparing them with those obtained by using a second (perhaps a reference) method. In this case we have two sample means $\bar{x}_1$ and $\bar{x}_2$. Taking the null hypothesis that the two methods give the same result, we need to test whether $(\bar{x}_1 - \bar{x}_2)$ differs significantly from zero. If the two samples have standard deviations which are not significantly different (see Section 3.5 for a method of testing this assumption), a **pooled** estimate of standard deviation can be calculated from the two individual standard deviations $s_1$ and $s_2$ by using the equation:

$$s^2 = \{(n_1 - 1)s_1^2 + (n_2 - 1)s_2^2\} / (n_1 + n_2 - 2) \tag{3.2}$$

It can be shown that $t$ is then given by:

$$t = (\bar{x}_1 - \bar{x}_2)/s \sqrt{(1/n_1 + 1/n_2)} \tag{3.3}$$

where $t$ has $(n_1 + n_2 - 2)$ degrees of freedom.

*Example.* In a comparison of two methods for the determination of boron in plant samples the following results ($\mu$g/g) were obtained:

spectrophotometric method:     mean = 28.0 ; standard deviation = 0.3
fluorimetric method:           mean = 26.25; standard deviation = 0.23

For each method 10 determinations were made.
(J. Anarez, A. Bonilla and J. C. Vidal, *Analyst,* 1983, **108**, 368).
Do these two methods give results having means which differ significantly?
The null hypothesis adopted is that the means of the results given by the two methods are equal. From Eq. (3.2), the pooled value of the standard deviation is

$$s^2 = (9 \times 0.3^2 + 9 \times 0.23^2)/18$$

$$= 0.0715$$

$$s = 0.267$$

From Eq. (3.3):

$$t = (28.0 - 26.25)/0.267 \sqrt{(1/10 + 1/10)}$$

$$= 14.7$$

There are 18 degrees of freedom, so (Table A.1) the critical value of $|t|$ ($P = 0.05$) is approximately 2.1: since the experimental value of $|t|$ is greater than this the difference between the two results is significant at the 5% level and the null hypothesis is rejected. In fact since the critical value of $|t|$ for $P = 0.001$ is about 3.9, the difference is significant even at the 0.1% level. In other words, if the null hypothesis is true, the probability of such a large difference arising by chance is less than 1 in 1000.

Another application of this test is illustrated by the following example where it is used to decide whether a change in the conditions of an experiment affects the result.

*Example.* In a series of experiments on the determination of tin in foodstuffs, samples were boiled with hydrochloric acid under reflux for different times. Some of the results are shown below:

| Refluxing time (min) | Tin found (mg/kg) |
|---|---|
| 30 | 55, 57, 59, 56, 56, 59 |
| 75 | 57, 55, 58, 59, 59, 59 |

(Analytical Methods Committee, *Analyst,* 1983, **108**, 109).
Does the mean amount of tin found differ significantly for the two boiling times?
The mean and variance (square of the standard deviation) for the two times are:

| | | |
|---|---|---|
| 30 min | $\bar{x}_1 = 57.0$ | $s_1^2 = 2.80$ |
| 75 min | $\bar{x}_2 = 57.8$ | $s_2^2 = 2.57$ |

The null hypothesis is adopted that boiling has no effect on the amount of tin found. From Eq. (3.2), the pooled value for the variance is

$$s^2 = (5 \times 2.80 + 5 \times 2.57)/10$$
$$= 2.685$$
$$s = 1.64$$

and $t$ is calculated from Eq. (3.3):

$$t = (57.0 - 57.8)/1.64 \sqrt{(1/6 + 1/6)}$$
$$= 0.84$$

There are 10 degrees of freedom so the critical value of $|t|$ is 2.23 ($P = 0.05$). The observed value of $|t|$ is less than the critical value so the null hypothesis is retained: there is no evidence that the length of boiling time affects the recovery rate.

If the assumption that the population standard deviations are equal is not valid, Eq. (3.3) is modified to:

$$t = (\bar{x}_1 - \bar{x}_2)/\sqrt{(s_1^2/n_1 + s_2^2/n_2)} \qquad (3.4)$$

and the number of degrees of freedom is calculated from:

$$\text{degrees of freedom} = \left\{ \frac{(s_1^2/n_1 + s_2^2/n_2)^2}{\dfrac{(s_1^2/n_1)^2}{n_1 + 1} + \dfrac{(s_2^2/n_2)^2}{n_2 + 1}} \right\} - 2 \qquad (3.5)$$

the result being rounded to the nearest whole number.

*Example.* The following table gives the concentration of thiol in the lysate of the blood of two groups of volunteers, the first group being 'normal' and the second suffering from rheumatoid arthritis.

Thiol concentration (m$M$)

| Normal | Rheumatoid |
|--------|------------|
| 1.84   | 2.81       |
| 1.92   | 4.06       |
| 1.94   | 3.62       |
| 1.92   | 3.27       |
| 1.85   | 3.27       |
| 1.91   | 3.76       |
| 2.07   |            |

(J. C. Banford, D. H. Brown, A. A. McConnell, C. J. McNeil, W. E. Smith, R. A. Hazelton and R. D. Sturrock, *Analyst,* 1983, **107**, 195).
From these data we have:

$$n_1 = 7, \ \bar{x}_1 = 1.921, \ s_1 = 0.076$$
$$n_2 = 6, \ \bar{x}_2 = 3.465, \ s_2 = 0.440$$

Again the null hypothesis adopted is that the mean concentration of thiol is the same for the two groups. Substitution in Eq. (3.4) gives $t = 8.5$ and from Eq. (3.5) there are 5 degrees of freedom. The critical value of $|t|$ ($P = 0.001$) is 6.87, so the null hypothesis is rejected: the mean concentration of thiol is different for the two groups.


## 3.4 PAIRED $t$-TEST

It frequently happens that two methods of analysis have to be compared by studying test samples containing substantially different amounts of analyte, as illustrated by the following example.

*Example.* The following table gives the concentration of lead ($\mu$g/l.) determined by two different methods for each of four test samples.

| Sample | Wet oxidation | Direct extraction |
|--------|---------------|-------------------|
| 1      | 71            | 76                |
| 2      | 61            | 68                |
| 3      | 50            | 48                |
| 4      | 60            | 57                |

(B. M. Smith and M. B. Griffiths, *Analyst,* 1982, **107**, 253).
Do the two methods give values for the mean lead concentration which differ significantly?

The test for comparing two means (Section 3.3) is not appropriate in this case because any variation due to method will be swamped by the effect of the differences between the test samples. This difficulty is overcome by looking at the difference between each pair of results given by the two methods. Adopting the null hypothesis that there is no significant difference in the mean concentrations given by the two methods, we can test whether the mean of the differences differs significantly from zero.

For the pairs of values above, the differences are $-5, -7, 2, 3$. The mean difference, $\bar{x}_d$, is $-1.75$ and the standard deviation of the differences, $s_d$, is 4.99. Since $\mu = 0$, Eq. (3.1) for calculating $t$ becomes:

$$t = \bar{x}_d \sqrt{n}/s_d \qquad\qquad (3.6)$$

where $t$ has $(n - 1)$ degrees of freedom. Substituting in Eq. (3.6) gives $t = -0.61$. The critical value of $|t|$ is 3.18 ($P = 0.05$) and since the calculated value of $|t|$ is less than this, the null hypothesis is retained: the methods do not give significantly different values for the mean lead concentration.

There are various circumstances in which a paired comparison is to be preferred to a comparison of means. Some examples are:

(a) the quantity of any one test sample available is sufficient for only one determination by each method;

(b) the methods are to be compared by using a wide variety of samples from different sources and possibly with different concentrations (but see the next paragraph);

(c) the test samples may be presented over an extended period so it is necessary to remove the effects of variations in the environmental conditions, such as temperature, pressure etc.

As analytical methods usually have to be applicable over a wide range of concentrations, a new method is often compared with a standard method by analysis of samples in which the analyte concentration may vary over several powers of ten. In this case it is inappropriate to use the paired *t*-test since its validity rests on the assumption that any errors, either random or systematic, are independent of concentration. Over wide ranges of concentration this assumption may no longer be true. The preferred statistical method in such cases is linear regression: its application to this situation is described in the next Chapter (Section 4.4).

## 3.5 *F*-TEST FOR THE COMPARISON OF STANDARD DEVIATIONS

The significance tests so far described in this chapter are used for comparing means, and hence for detecting systematic errors. It is also important in many cases to compare the standard deviations, i.e. the random errors of two sets of data. This comparison can take two forms. Either we may wish to test whether Method A is more precise than Method B (known as a **one-tailed test**) or we may wish to test whether Methods A and B differ in their precision (known as a **two-tailed test**). The difference between these two situations may not be clear at first sight, but in practice is very important. In the first case we assume that there is no possibility of Method B being more precise than Method A − we only wish to test the converse possibility. In the second case we have an entirely open mind about the relative precisions of the two methods, and wish to test for any significant difference. Thus, if we wished to test whether a new analytical method was *more* precise than a standard method we would use a one-tailed test: if we wished to test whether two standard deviations differed significantly (e.g. before carrying out a *t*-test − see Section 3.3) a two-tailed test would be appropriate.

The *F*-**test** considers the ratio of the two sample variances, i.e. the ratio of the squares of the standard deviations. The quantity calculated ($F$) is given by:

$$F = s_1^2 / s_2^2 \qquad\qquad (3.7)$$

$s_1^2$ and $s_2^2$ being allocated in the equation so that $F$ is always $\geqslant 1$. The null hypothesis adopted is that the populations from which the samples are taken are normal, and that the population variances are equal. If the null hypothesis is true then the variance ratio should be close to 1. Differences from 1 occur because of random variation but if the difference is too great it can no longer be attributed to this cause: if the calculated value of $F$ exceeds a certain value (obtained from tables) then the null hypothesis is rejected. This critical value of $F$ depends on the size of the two samples, the significance level and the type of test performed. The values for $P = 0.05$ are given in Tables A.2 and A.3; the use of the Tables is illustrated in the following examples.

*Example.* A proposed method for the determination of the chemical oxygen demand of wastewater was compared with the standard (mercury salts) method. The following results were obtained for a sewage effluent sample.

|                   | Mean (mg/l.) | Standard deviation (mg/l.) |
|-------------------|--------------|----------------------------|
| Standard method   | 72           | 3.31                       |
| Proposed method   | 72           | 1.51                       |

For each method 8 determinations were made.
(D. Ballinger, A. Lloyd and A. Morrish, *Analyst,* 1982, **107**, 1047).
Is the precision of the proposed method significantly greater than that of the standard method?

We have to decide whether the variance of the standard method is significantly greater than that of the proposed method. *F* is given by the ratio of the variances:

$$F_{7,7} = 3.31^2 / 1.51^2 = 4.8$$

The samples both comprise 8 values, so the number of degrees of freedom in each case is 7, as indicated by the subscripts. This is a case where a one-tailed test must be used, the only point of interest being whether the proposed method is *more* precise than the standard method. In Table A.2 the number of degrees of freedom of the denominator is given in the left-hand column and the number of degrees of freedom of the numerator at the top. The critical value of $F$ in this case is 3.787 ($P = 0.05$). Since the calculated value of $F$ (4.8) exceeds this, the variance of the standard method *is* significantly greater than that of the proposed method at the 5% probability level, i.e. the proposed method is the more precise.

*Example.* In the example at the beginning of Section 3.3 it was assumed that the variances of the two methods for determining boron in plants did not differ significantly. This assumption can now be tested. The standard deviations were 0.3 and 0.23 (each obtained from 10 measurements on a specimen of a particular plant). To calculate $F$ so that it is $\geqslant 1$ we must write

$$F_{9,9} = 0.3^2 / 0.23^2 = 1.7$$

In this case, however, we have no reason to expect in advance that the variance of one method should be greater than the other, so a two-tailed test is appropriate. The critical values given in Table A.2 are the values that $F$ will exceed, with a probability of 0.05, *assuming that it must be greater than 1*. In a two-tailed test the ratio of the first to the second variance could be less or greater than 1 but, if $F$ is always calculated so that it is *greater* than 1, then the probability that it exceeds the critical values given in Table A.2 will be doubled. Thus these critical values are not appropriate for a two-tailed test and Table A.3 is used instead. From this Table, taking the number of degrees of freedom of both numerator and denominator as 9, the critical value of $F$ is 4.026. The calculated value is less than this, so there is no significant difference between the two variances at the 5% level.

As with the $t$-test, other significance levels may be used for the $F$-test and the critical values can be found from the tables listed in the bibliography at the end of this chapter. Care must be taken that the correct table is used, depending on whether the test is one- or two-tailed: for an $\alpha$% significance level the $\alpha$% points of the $F$-distribution are used for a one-tailed test and the $\frac{1}{2}\alpha$% points are used for a two-tailed test.

## 3.6 OUTLIERS

Every experimentalist is familiar with the situation in which one (or possibly more) of a set of results appears to differ unreasonably from the others in the set. Such a measurement is called an **outlier**. In some cases an outlier may be attributable to human error. For example if the following results were given for replicate titrations:

$$12.12, \ 12.15, \ 12.13, \ 13.14, \ 12.12 \text{ ml}$$

then the fourth value is almost certainly due to a slip in writing down the result and should read 12.14. However, even when such obviously erroneous values have been removed or corrected, values which appear to be outliers may still occur. Should they be kept, come what may, or should some means be found to test statistically whether or not they should be rejected? Obviously the final values presented for the mean and standard deviation will depend on whether or not the outliers are rejected. Since discussion of the precision and accuracy of a method depends on these final values, it should always be made clear whether outliers have been rejected, and if so, why.

The discussion of errors so far has rested on the assumption that the distribution of repeated measurements is normal or nearly so. One reason for the presence of outliers may be that this assumption is not valid: we assume a particular, i.e. normal, model for the population from which the measurements are drawn and because this model is not correct we obtain a result which surprises us. This section will describe only outlier tests which assume that the parent population is normal. If this assumption is not correct then the tests described so far in this chapter should not be used (unless the samples are large, in which case the sampling distributions become normal) but should be replaced by the appropriate non-parametric methods described in Chapter 5. The latter are generally insensitive to extreme values and so the problem of whether or not outliers should be rejected is avoided.

One way (amongst several different tests) of assessing a suspect measurement is to compare the difference between it and the measurement nearest to it in size with the difference between the highest and lowest measurements. The ratio of these differences (without regard to sign) is known as **Dixon's** $Q$.

$$Q = |\text{suspect value} - \text{nearest value}|/(\text{largest value} - \text{smallest value}) \qquad (3.8)$$

The critical values of $Q$ for $P = 0.05$ and $P = 0.01$ are given in Table A.4. If the calculated value of $Q$ exceeds the critical value the suspect value is rejected.

*Example.* The following values were obtained for the nitrite concentration (mg/l.) in a sample of river water:

$$0.403, \; 0.410, \; 0.401, \; 0.380$$

The last measurement is suspect: should it be rejected?
We have:

$$Q = |0.380 - 0.401|/(0.410 - 0.380) = 0.021/0.030 = 0.7$$

From Table A.4, for sample size 4, the critical value of $Q$ is $0.831$ ($P = 0.05$). Since the calculated value of $Q$ does not exceed this, the suspect measurement should be retained.

Ideally, further measurements should be made when a suspect value occurs, particularly if only a few values have been obtained initially. This may make it clearer whether or not the suspect value should be rejected, and if it is still retained will also reduce to some extent its effect on the mean and standard deviation.

*Example.* If three further measurements were added to those given in the example above, so that the complete results became:

$$0.403, \ 0.410, \ 0.401, \ 0.380, \ 0.400, \ 0.413, \ 0.411$$

should 0.380 still be retained?

The calculated value of $Q$ is now:

$$Q = |0.380 - 0.400|/(0.413 - 0.380) = 0.606$$

The critical value of $Q$ ($P = 0.05$) for sample size 7 is 0.570, so the suspect measurement is rejected at the 5% significance level.

It is important to appreciate that for a significance level of 5% there is still a 5% chance, or 1 in 20, of incorrectly rejecting the suspect value. This may have a considerable effect on the estimation of the precision of an experiment. For example, for all 7 values of the nitrite concentration given above, the standard deviation is 0.011 mg/l. but when the suspect value is rejected the standard deviation becomes 0.0056 mg/l., i.e. the precision appears to have improved by a factor of 2.

The example above illustrates the importance of caution in rejecting outliers. When measurements are repeated only a few times (which is common in analytical work), rejection of one value makes a great difference to the mean and standard deviation. In particular, the practice of making three measurements and rejecting the one which differs most from the other two should be avoided, even if the $Q$ test indicates that one value should be rejected. It can be shown that a more reliable estimate of the mean is obtained, on average, by using the middle one of the three values rather than the mean of the two unrejected values.

As well as the test for one outlier described above, there are tests for two outliers. Without pursuing these in detail, we can see that two suspect values, either close to each other or one at each end of the range of values, will reduce the calculated value of $Q$. In the first case there will be a reduction in the numerator and in the second case an increase in the denominator of the fraction. In either case the effect of the most extreme measurement will be **masked** in the $Q$-test by the presence of the other possible outlier. This effect can be seen by considering the following values:

$$2.1, \ 2.0, \ 2.1, \ 2.3, \ 2.9, \ 2.3, \ 3.1, \ 2.2, \ 2.0, \ 2.3$$

Obviously 2.9 and 3.1 are suspect values, but the calculated value of $Q$ is:

$$Q = (3.1 - 2.9)/(3.1 - 2.0) = 0.18$$

a value which is not significant ($P = 0.05$). In such cases a test for an outlier pair is appropriate and references to such tests are given in the bibliography at the end of the chapter.

Outliers also occur in other statistical calculations such as the linear regression methods described in Chapter 4. These cases are also discussed in advanced texts.

## 3.7 CONTROL CHARTS

One important application of significance testing is in the field of quality control, where a process is monitored periodically in order to test for any changes in performance. A particular example would be the situation in which small samples are taken at regular intervals in order to test the average quality of items produced by a particular process. The size of sample taken is normally about 4-6. (Sample size is discussed later in this section and also in Chapter 6, where methods of taking samples are further considered.) Ideally all the items produced should conform to a target value, $\mu$, but in practice there will be some random variation from one item to the next. The size of this random variation, as measured by the population standard deviation, $\sigma$, is usually known from past experience. If the sample size is $n$, we know from Chapter 2 [Eqs. (2.5) and (2.6)] that approximately 95% of the sample means should lie within the range $\mu \pm 2(\sigma/\sqrt{n})$ and approximately 99.7% of the sample means within the range $\mu \pm 3(\sigma/\sqrt{n})$ providing that the process mean is unchanged, i.e. the process is under control. The object of a **control chart** is to present the values for the sample means graphically in such a way that any corrective action can be taken as quickly as possible. Figure 3.1 shows a type of control chart, known as a **Shewhart chart**,

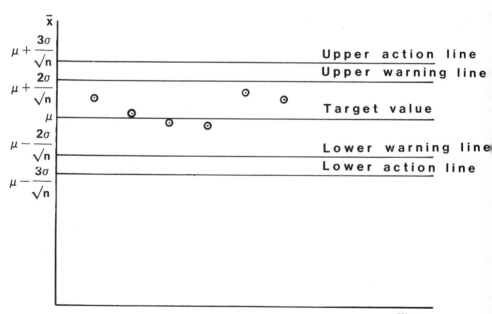

Fig. 3.1 – A typical Shewhart chart.

for the process described above. The value for the sample mean, $\bar{x}$, is plotted against time, and when the process is under control the values of $\bar{x}$ are normally distributed about $\mu$. There are also two pairs of horizontal lines on the chart: the **warning lines** at $\pm 2\sigma/\sqrt{n}$ and the **action lines** at $\pm 3\sigma/\sqrt{n}$. The purpose of these lines is indicated by their names. Since the probability that a sample mean falls outside the action lines, when the process is under control, is only 0.003 (i.e. 3 in 1000), the process is usually stopped and investigated if this happens. The probability of $\bar{x}$ falling between the action and warning lines is about 0.05 (i.e. 1 in 40): one such point would not give cause for concern but if two such points are obtained consecutively then the process should be stopped. It can be seen that a control chart displays a series of significance tests with the warning and action lines corresponding to the critical values for $P = 0.05$ and $P = 0.003$ respectively.

It is usually important to monitor the standard deviation of a process as well as its mean value. This is most easily done by using the range $w$, (i.e. the difference between the highest and lowest values) of each sample taken. Usually only an increase in variability is of importance and so only the upper action and warning lines in the control chart for the standard deviation are of interest. The appropriate control chart can be constructed with the aid of tables which give the action and warning lines, and the target value of $w$ for different values of $n$ and $\sigma$ (see bibliography to this chapter).

An important consideration in using a Shewhart chart is how quickly a change in the process mean can be detected. For example, if the process mean changes by $3\sigma/\sqrt{n}$ then there is a probability of $1/2$ that the next point will fall outside the action lines; if the process mean changes by $1\sigma/\sqrt{n}$ this probability falls to $1/40$. The average number of points which have to be plotted before a change in the process mean is detected is known as the **average run length (A.R.L.)**. Obviously it depends on the size of the change in the process mean, compared with $\sigma/\sqrt{n}$: the greater the change the more quickly it is detected. For example, if the process mean changes by $1\sigma/\sqrt{n}$, then the A.R.L. before a sample mean falls outside the action lines is about 50. If the process is also stopped when two consecutive points fall outside the warning lines, the A.R.L. is approximately halved. Since the A.R.L. depends on the change in $\mu$ compared with $\sigma/\sqrt{n}$, it can be reduced by increasing the sample size: tne largest possible sample size is usually dictated by considerations of cost and time. The A.R.L. can be further reduced, for a particular sample size, by using a different type of control chart, known as a **cusum chart**. This utilizes all the previous sample means rather than just the last one or two as a Shewhart chart does. 'Cusum' is an abbreviation for 'cumulative sum', i.e. the sum of the deviations of the sample means from the target value, carried forward cumulatively. An example should make its calculation clear.

*Example.* Table 3.1 gives the value of the mean for a sequence of samples.

The value of $\sigma/\sqrt{n}$ is known to be 2.5 and Fig. 3.2 shows the Shewhart chart for the sample means. It can be seen that although no points fall outside the warning lines, a sequence of values falls on one side of the target value.

**Table 3.1** – Calculation of cumulative sum

| Target value = 80 | | $\sigma/\sqrt{n}$ (known) = 2.5 | |
|---|---|---|---|
| Observation number | Sample mean | Sample mean – target value | Cusum |
| 1 | 82 | +2 | +2 |
| 2 | 79 | −1 | +1 |
| 3 | 80 | 0 | +1 |
| 4 | 78 | −2 | −1 |
| 5 | 82 | +2 | +1 |
| 6 | 79 | −1 | 0 |
| 7 | 80 | 0 | 0 |
| 8 | 79 | −1 | −1 |
| 9 | 78 | −2 | −3 |
| 10 | 80 | 0 | −3 |
| 11 | 76 | −4 | −7 |
| 12 | 77 | −3 | −10 |
| 13 | 76 | −4 | −14 |
| 14 | 76 | −4 | −18 |
| 15 | 75 | −5 | −23 |

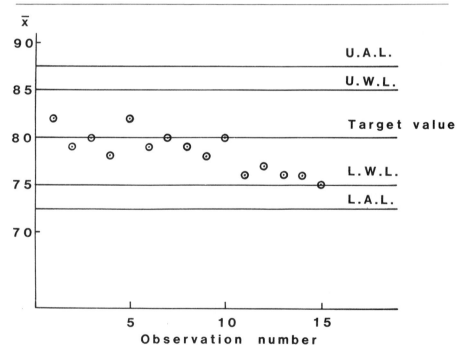

Fig. 3.2 – Shewhart chart for the data of Table 3.1 (U.A.L. stands for upper action line, and so on).

Table 3.1 also shows the calculation of the cusum. Obviously, if the process is under control, positive and negative deviations are equally likely, so the cusum should oscillate about zero. The values of the cusum are plotted in Fig. 3.3. To give good visual impact the cusum chart is drawn with the distance corresponding to $2\sigma/\sqrt{n}$ on the vertical axis equal to the distance between successive observations on the horizontal axis.

From the cusum chart it appears that the process mean changed after the eighth sample was taken. This is one advantage of the cusum chart – it indicates at what point the process went out of control.

In order to test whether a trend in a cusum chart *does* indicate that the process mean has changed and that the change cannot be accounted for simply by random variation, a **V-mask** is used. As shown in Fig. 3.4, a V-shaped mask, preferably engraved on perspex, is placed on the cusum chart with its axis horizontal and its apex a distance $d$ to the right of the last observation. The angle between the arms of the V is denoted by $2\theta$. The process is taken to be under control if all the values of the cusum fall between the arms of the V as shown in Fig. 3.4. Figure 3.5 illustrates a situation in which the process is out of control: two of the cusum values are outside the upper arm of the V-mask, which indicates that the process mean has fallen below the target value (see Figs. 3.2 and 3.3). Obviously the performance of the mask depends on the values taken for $\theta$ and $d$. These values need to be chosen so that very few false alarms occur when the process is under control, but an important change in the

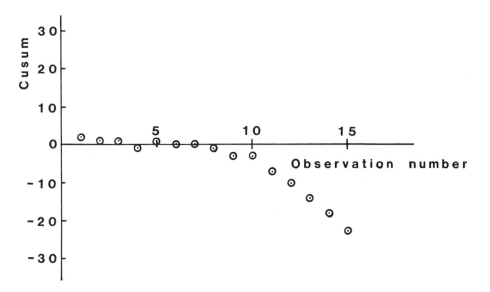

Fig. 3.3 – Cusum chart for the data of Table 3.1.

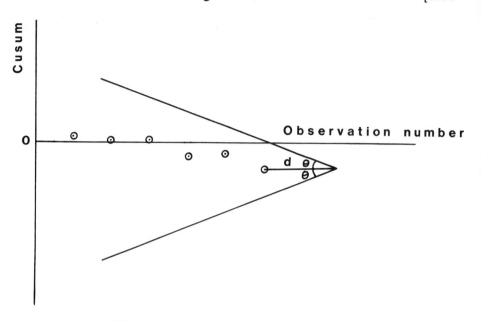

Fig. 3.4 – Use of a V-mask with the process in control.

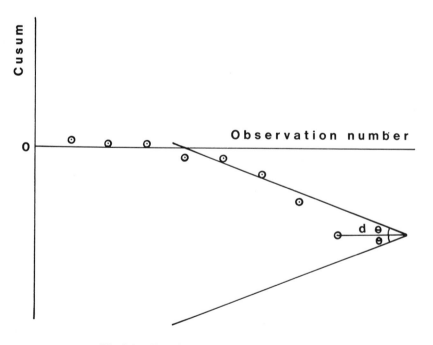

FIg. 3.5 – Use of a V-mask with the process out of control.

process mean is quickly detected. Further details of the dependence of the A.R.L. on $d$ and $\theta$ for different changes in the process mean are given in an I.C.I. monograph cited in the bibliography to this chapter.

A cusum chart can also be used to estimate the size of the change which has occurred in the process mean when the process has gone out of control. If, for example, the process mean decreases by $\Delta$, then on average the sample means will be $\Delta$ less than the target value. As a result the cusum will decrease on average by $\Delta$ for each point plotted. Thus the average slope of the line joining the cusum points gives a measure of the correction needed.

## 3.8 ANALYSIS OF VARIANCE

In Section 3.3 a method was described for comparing two means to test whether they differed significantly. In analytical work there are often more than two means to be compared. Some possible situations are: comparing the mean concentration of protein in a solution for samples stored under different conditions; comparing the mean results obtained for the concentration of an analyte by several different methods; comparing the mean titration results obtained by several different experimentalists using the same apparatus. In all these examples there are two possible sources of variation. The first, which is always present, is due to random error. This was discussed in detail in the previous chapter: it is this error which causes different results to be obtained when measurements are repeated under the same conditions. The second possible source of variation is due to what is known as a **controlled** or **fixed-effect factor**: for the examples above, the controlled factors are, respectively, the conditions under which the solution was stored, the method of analysis used, and the experimentalist carrying out the titration. **Analysis of variance** (frequently abbreviated to **ANOVA**) is an extremely powerful statistical technique which can be used to separate and estimate the different causes of variation. For the particular examples above, it can be used to separate any variation which is caused by changing the controlled factor from the variation due to random error. It can thus test whether altering the controlled factor leads to a significant difference between the mean values obtained.

ANOVA can also be used in situations where there is more than one source of random variation. Consider, for example, the purity testing of a barrelful of sodium chloride. Samples are taken from different parts of the barrel, chosen at random, and replicate analyses are performed on these samples. In addition to the random error in the measurement of the purity, there may also be variation in the purity of the samples from different parts of the barrel. Since the samples were chosen at random, this variation will be random and is thus sometimes known as a **random-effect** or **uncontrolled factor**. Again, ANOVA can be used to separate and estimate the sources of variation.

Both types of statistical analysis described above, i.e. where there is *one* factor, either controlled or random, in addition to the random error, are known as **one-way** **ANOVA**. The arithmetical procedures are similar in the fixed- and random-effect factor cases and examples of both will be given below. More complex situations in which there are two or more factors, possibly interacting with each other, are considered in Chapter 6 (Experimental Design).

### 3.9 COMPARISON OF SEVERAL MEANS

Table 3.2 shows the results obtained in an investigation into the stability of a fluorescent reagent stored under different conditions. The values given are the fluorescence signals (in arbitary units) from dilute solutions of equal concentration. Three replicate measurements were made on each sample.

**Table 3.2** – Fluorescence signal from solutions stored under different conditions

| Conditions | Replicate Measurements | Mean |
|---|---|---|
| A Freshly prepared | 102, 100, 101 | 101 |
| B Stored for 1 hr in the dark | 101, 101, 104 | 102 |
| C Stored for 1 hr in subdued light | 97, 95, 99 | 97 |
| D Stored for 1 hr in bright light | 90, 92, 94 | 92 |
| | Overall mean | 98 |

The table shows that the mean values for the four samples are different. However, we know that because of random error, even if the true value which we are trying to measure is unchanged, the sample mean may vary from one sample to the next. ANOVA tests whether the difference between the sample means is too great to be explained by the random error.

The problem can be generalized by considering $h$ samples, each with $n$ members, as in Table 3.3 where $x_{ij}$ is the $j$th measurement on the $i$th sample. The means of the samples are $\bar{x}_1, \bar{x}_2, \ldots \bar{x}_h$ and the mean of all the values grouped together is $\bar{x}$.

**Table 3.3** – Generalization of Table 3.2

| | | | | | Mean |
|---|---|---|---|---|---|
| Sample 1 | $x_{11}$ | $x_{12}$ $\cdots\cdots$ $x_{1j}$ | $\cdots\cdots$ $x_{1n}$ | | $\bar{x}_1$ |
| Sample 2 | $x_{21}$ | $x_{22}$ $\cdots\cdots$ $x_{2j}$ | $\cdots\cdots$ $x_{2n}$ | | $\bar{x}_2$ |
| Sample $i$ | $x_{i1}$ | $x_{i2}$ $\cdots\cdots$ $x_{ij}$ | $\cdots\cdots$ $x_{in}$ | | $\bar{x}_i$ |
| Sample $h$ | $x_{h1}$ | $x_{h2}$ $\cdots\cdots$ $x_{hj}$ | $\cdots\cdots$ $x_{hn}$ | | $\bar{x}_h$ |
| | | | | overall mean $= \bar{x}$ | |

The null hypothesis adopted is that all the samples are drawn from a population with mean $\mu$ and variance $\sigma_0^2$. On the basis of this hypothesis $\sigma_0^2$ can be estimated in two ways, one involving the variation *within* the samples and the other the variation *between* the samples.

(i) *Within-sample variation*
For each sample a variance can be calculated by using the formula $\Sigma(x - \bar{x})^2 /$ $(n - 1)$ [see Eq. (2.2)] . Using the values in Table 3.2 we have:

$$\text{Variance of sample A} = \frac{(102 - 101)^2 + (100 - 101)^2 + (101 - 101)^2}{3 - 1}$$

$$= 1$$

$$\text{Variance of sample B} = \frac{(101 - 102)^2 + (101 - 102)^2 + (104 - 102)^2}{3 - 1}$$

$$= 3$$

$$\text{Variance of sample C} = \frac{(97 - 97)^2 + (95 - 97)^2 + (99 - 97)^2}{3 - 1}$$

$$= 4$$

$$\text{Variance of sample D} = \frac{(89 - 91)^2 + (91 - 91)^2 + (93 - 91)^2}{3 - 1}$$

$$= 4$$

Averaging these values gives:

$$\text{within-sample estimate of } \sigma_0^2 = (1 + 3 + 4 + 4)/4 = 3$$

This estimate has 8 degrees of freedom: each sample estimate has 2 degrees of freedom and there are 4 samples. Note that this estimate does not depend on the means of the samples: for example, if all the measurements for sample A were increased by say, 4, this estimate of $\sigma_0^2$ would be unaltered.

The general formula for the within-sample estimate of $\sigma_0^2$ is:

$$\text{within-sample estimate of } \sigma_0^2 = \underset{i\ j}{\Sigma\ \Sigma}\ (x_{ij} - \bar{x}_i)^2 / h(n - 1) \qquad (3.9)$$

The summation over $j$ measurements and division by $(n - 1)$ gives the variance of each sample; the summation over $i$ samples and division by $h$ averages these sample variances. The expression in Eq. (3.9) is known as a *mean square* since it involves a sum of squared terms divided by the number of degrees of freedom. Since in this case the number of degrees of freedom is 8 and the mean square is 3, the sum of the squared terms is $3 \times 8 = 24$.

(ii) *Between-sample variation*
If the samples are all drawn from a population which has variance $\sigma_0^2$, then their

means come from a population with variance $\sigma_0^2/n$ (cf. the sampling distribution of the mean, Section 2.3). Thus, if the null hypothesis is true, the variance of the means of the samples gives an estimate of $\sigma_0^2/n$. From Table 3.2:

$$\text{sample-mean variance} = \frac{(101-98)^2 + (102-98)^2 + (97-98)^2 + (92-98)^2}{4-1}$$

$$= 62/3$$

so

$$\text{between-sample estimate of } \sigma_0^2 = \frac{62}{3} \times 3 = 62$$

This estimate has 3 degrees of freedom since it is calculated from 4 sample means. Note that this estimate of $\sigma_0^2$ does not depend on the variability *within* each sample, since it is calculated from the sample means. However, if, for example, the mean of sample $D$ were changed, then this estimate of $\sigma_0^2$ would also be changed.

In general we have:

$$\text{between-sample estimate of } \sigma_0^2 = n \sum_i (\bar{x}_i - \bar{x})^2/(h-1) \qquad (3.10)$$

which again is a 'mean square' involving a sum of squared terms divided by the number of degrees of freedom. In this case the number of degrees of freedom is 3 and the mean square is 62 so the sum of the squared terms is $3 \times 62 = 186$.

Summarizing our calculations so far:

within-sample mean square  $= 3$, with 8 d.f.
between-sample mean square $= 62$ with 3 d.f.

If the null hypothesis is correct, these two estimates of $\sigma_0^2$ should not differ significantly. If it is incorrect, the between-sample estimate of $\sigma_0^2$ will be greater than the within-sample estimate because of between-sample variation. To test whether the between-sample estimate is significantly greater a one-tailed $F$-test is used (see Section 3.5):

$$F_{3,8} = 62/3 = 20.7$$

(Remember each mean *square* is used, so no further squaring is necessary.) From Table A.2 the critical value of $F$ is 4.066 ($P = 0.05$). Since the calculated value of $F$ is greater than this, the null hypothesis is rejected: the sample means do differ significantly.

A significant result in one-way ANOVA can arise for several different reasons: for example, one mean may differ from all the others, all the means may differ from each other, the means may fall into two distinct groups etc. A simple way of deciding the reason for a significant result is to arrange the means in order of increasing size and compare the difference between adjacent values with a quantity called the **least significant difference**. This is given by $s\sqrt{(2/n)} \times t_{h(n-1)}$ where $s$

is the within-sample estimate of $\sigma_0$ and $h(n - 1)$ is the number of degrees of freedom of this estimate. For the example above, the sample means arranged in increasing order of size are:

$$\bar{x}_D = 92; \ \bar{x}_C = 97; \ \bar{x}_A = 101; \ \bar{x}_B = 102$$

and the least significant difference is $\sqrt{3} \times \sqrt{(2/3)} \times 2.306$ ($P = 0.05$), giving 3.26. Comparing this value with the differences between the means suggests that $\bar{x}_D$ and $\bar{x}_C$ differ significantly from each other and from $\bar{x}_A$ and $\bar{x}_B$ but that $\bar{x}_A$ and $\bar{x}_B$ do not differ signficantly from each other, i.e. it is exposure to light which affects the fluorescence.

The least significance difference method described above is not entirely rigorous: it can be shown that it leads to rather too many significant differences. However it is a simple follow-up test when ANOVA has indicated that there *is* a significant difference between the means. Descriptions of other more rigorous tests are given in the references at the end of this chapter.

## 3.10 THE ARITHMETIC OF ANOVA CALCULATIONS

In using ANOVA to test for the difference between several means, $\sigma_0^2$ was estimated in two different ways. If the null hypothesis were true, $\sigma_0^2$ could also be estimated in a third way by treating the data as one large sample. This would involve summing the squares of the deviations from the overall mean:

$$\sum_i \sum_j (x_{ij} - \bar{x})^2 = 4^2 + 2^2 + 3^2 + 3^2 + 3^2 + 6^2 + 1^2 + 3^2 + 1^2 + 8^2 + 6^2 + 4^2$$
$$= 210$$

and dividing by the number of degrees of freedom, $12 - 1 = 11$.

This method of estimating $\sigma_0^2$ is not used in the analysis because the estimate depends both on the within- and between-sample variation. However, there is an exact algebraic relationship between this total variation and sources of variation which contribute to it, which, especially in more complicated ANOVA calculations, leads to a simplification of the arithmetic involved. The relationship between the sources of variation is illustrated by Table 3.4 which summarizes the sums of squares and degrees of freedom.

**Table 3.4** – Summary of sums of squares and degrees of freedom

| Source of variation | Sum of squares | Degrees of freedom |
|---|---|---|
| Between-sample | $n \sum_i (\bar{x}_i - \bar{x})^2 = 186$ | $h - 1 = 3$ |
| Within-sample | $\sum_i \sum_j (x_{ij} - \bar{x}_i)^2 = 24$ | $h(n - 1) = 8$ |
| Total | $\sum_i \sum_j (x_{ij} - \bar{x})^2 = 210$ | $hn - 1 = 11$ |

It will be seen that the values for the total variation, given in the last row of the table, are the sums of the values in the first two rows for both the sum of squares and the degrees of freedom. This additive property holds for all the ANOVA calculations described in this book.

Just as in the calculation of variance, there are formulae which simplify the calculation of the individual sums of squares. These formulae are summarized in Table 3.5 which uses the notation above and also introduces the symbols:

Total number of measurements $= N = nh$
Sum of the measurements in the $i$th sample $= T_i$
Sum of all the measurements, grand total $= T$

The additive property described above is used for the calculation of the within-sample sum of squares and degrees of freedom.

**Table 3.5** – Formulae for one-way ANOVA calculations

| Source of variation | Sum of squares | Degrees of freedom |
|---|---|---|
| Between-sample | $\sum_i T_i^2/n - T^2/N$ | $h - 1$ |
| Within-sample | by subtraction | by subtraction |
| Total | $\sum_i \sum_j x_{ij}^2 - T^2/N$ | $N - 1$ |

The use of these formulae is illustrated by a calculation in Section 3.11.

Certain assumptions have been made in performing the ANOVA calculations above. The first is that the variance of the random error is not affected by the treatment used. This assumption is implicit in the pooling of the within-sample variances to calculate an overall estimate of the error variance. In doing this we are assuming what is known as the **homogeneity of variance**. In the particular example given above, where all the measurements are made in the same way, we would expect homogeneity of variance. Various methods of testing for this property are given in the references at the end of this chapter.

A second assumption is that the uncontrolled variation is random. This would not be the case if, for example, there were some uncontrolled factor, such as temperature change, which produced a trend in the results over a period of time. The effect of such uncontrolled factors can be overcome to a large extent by the techniques of randomization and blocking which are discussed in Chapter 6.

It will be seen that an important part of ANOVA is the application of the $F$-test. Use of this test (cf. Section 3.5) simply to compare the variances of two samples, depends on the samples being drawn from a normal population. Fortunately, however, the $F$-test as applied in ANOVA is not too sensitive to departures from normality of distribution.

## 3.11  SEPARATION AND ESTIMATION OF VARIANCES BY USING ANOVA

In this section we consider the random-effect factor situation described in Section 3.8. In this case the purpose of ANOVA is not to test whether several sample means differ significantly, but to separate and estimate the different sources of variation. Table 3.7 shows the results of the purity testing of a barrelful of sodium chloride. Five samples were taken from different parts of the barrel, chosen at random, and four replicate analyses were performed on each sample.

**Table 3.7** – Purity testing of sodium chloride (% purity)

| Sample | Purity (%) | Mean |
|--------|------------|------|
| A | 98.8, 98.7, 98.9, 98.8 | 98.8 |
| B | 99.3, 98.7, 98.8, 99.2 | 99.0 |
| C | 98.3, 98.5, 98.8, 98.8 | 98.6 |
| D | 98.0, 97.7, 97.4, 97.3 | 97.6 |
| E | 99.3, 99.4, 99.9, 99.4 | 99.5 |

As explained in Section 3.8, there are two possible sources of variation. The first is the random error in the measurement of purity. We shall assume that this is normally distributed, with variance $\sigma_0^2$. The second is variation in the purity at different points in the barrel. We shall assume that this is normally distributed, with variance $\sigma_1^2$. Since the within-sample mean square does not depend on the sample mean (see Section 3.9) it can be used to give an estimate of $\sigma_0^2$. The between-sample mean square *cannot* be used to estimate $\sigma_1^2$ directly, because the variation between sample means is caused both by the random error in measurement and by possible variation in the purity. It can be shown that the between-sample mean square gives an estimate of $\sigma_0^2 + n\sigma_1^2$ (where, as before, $n$ is the number of replicate measurements).

Before an estimate of $\sigma_1^2$ is made, a test should be carried out to see whether it differs significantly from zero. If $\sigma_1^2 = 0$ then the within- and between-sample mean squares should not differ significantly, since in this case they both estimate $\sigma_0^2$. The calculation of the mean squares by using the formulae in Table 3.7 is set out below. All the values in Table 3.7 have had 98.5 subtracted from them, which simplifies the arithmetic considerably. Note that this does not affect either the within- or the between-sample estimates of variance, because the same quantity has been subtracted from *every* value.

| | | | | | $T_i$ | $T_i^2$ |
|---|---|---|---|---|---|---|
| Sample A | 0.3 | 0.2 | 0.4 | 0.3 | 1.2 | 1.44 |
| Sample B | 0.8 | 0.2 | 0.3 | 0.7 | 2.0 | 4.00 |
| Sample C | 0.2 | 0.0 | 0.3 | 0.3 | 0.4 | 0.16 |
| Sample D | −0.5 | −0.8 | −1.1 | −1.2 | −3.6 | 12.96 |
| Sample E | 0.8 | 0.9 | 1.4 | 0.9 | 4.0 | 16.00 |

$$T = 4.0 \quad \sum_i T_i^2 = 34.56$$

$n = 4;\ h = 5;\ N = 20;\ \sum_i \sum_j x_{ij}^2 = 9.62$

| Source of variation | Sum of squares | Degrees of freedom | Mean square |
|---|---|---|---|
| Between-sample | $34.56/4 - 4.0^2/20 = 7.84$ | 4 | $7.8/4 = 1.96$ |
| Within-sample | by subtraction $= 0.98$ | 15 | $0.98/15 = 0.0653$ |
| Total | $9.62 - 4.0^2/20 = 8.82$ | 19 | |

Since the between-sample mean square is greater than the within-sample mean square, $\sigma_1^2$ *may* differ significantly from 0. Using the $F$-test to compare the two mean squares gives:

$$F_{4,15} = 1.96/0.0653 = 30$$

From Table A.2 the critical value of $F$ is 3.056 ($P = 0.05$). The calculated value of $F$ exceeds this and so $\sigma_1^2$ differs significantly from 0. The within-sample mean square gives 0.0653 as an estimate of $\sigma_0^2$. Since the between-sample mean square estimates $\sigma_0^2 + n\sigma_1^2$ we have:

estimate of $\sigma_1^2 =$ (between-sample mean square $-$ within-sample mean
    square)$/n$
  $= (1.96 - 0.0653)/4$
  $= 0.47$

A similar analysis can be used to separate and estimate the different sources of error which cause 'between-run' and 'within-run' variation. These were discussed in some detail in Section 1.3. In general the between-run variation is greater than the within-run variation because of additional uncontrolled sources of variation in the former case. This additional between-run variation is analogous to the variation in the purity in the example given in this section.

## 3.12 THE CHI-SQUARED TEST

The significance tests so far described in this chapter have, in general, been concerned with testing whether the mean of several observations differs significantly from the value proposed by the null hypothesis. The data used have taken the form of observations which, apart from any rounding off, have been measured on a continuous scale. In contrast, this section is concerned with *frequencies*, i.e. the number of times a given event occurs. For example, Table 2.3 gives the frequencies of the different values obtained for the nitrate ion concentration when 50 measurements were made on a sample. As discussed in Chapter 2, such measurements are usually assumed to be drawn from a population which is normally distributed: the chi-squared test can be used to test whether the observed frequencies differ significantly from those which would be expected on this null hypothesis. Since the calculation involved in this case is relatively complicated it will not be described here. (A reference to a worked example is given at the end of the chapter.) The principle of the chi-squared test is more easily understood by means of the following example.

*Example.* The numbers of glassware breakages reported by four laboratory workers over a given period are shown below. Is there any evidence that the workers differ in their reliability?

Number of breakages: 24, 17, 11, 9

The null hypothesis adopted is that there is no difference in reliability. Assuming that the workers use the laboratory for an equal time, we would expect, from the null hypothesis, the same number of breakages for each worker. Since the total number of breakages is 61, the expected number of breakages per worker is $61/4 = 15.25$. Obviously it is not possible in practice to have a non-integral number of breakages: this number is a mathematical concept. The nearest practicable 'equal' distribution is 15, 15, 15, 16, in some order. The question to be answered is whether the difference between the observed and expected frequencies is so large that the null hypothesis should be rejected. That there should be *some* difference between the two sets of frequencies can be most readily appreciated by considering a sequence of throws of a die: we should, for example, be most surprised if 30 throws yielded exactly equal frequencies for 1, 2, 3 etc. The calculation of chi-squared, $\chi^2$, the quantity used to test for a significant difference, is shown below.

| Observed frequency, $O$ | Expected frequency, $E$ | $O - E$ | $(O - E)^2/E$ |
|:---:|:---:|:---:|:---:|
| 24 | 15.25 | 8.75 | 5.020 |
| 17 | 15.25 | 1.75 | 0.201 |
| 11 | 15.25 | −4.25 | 1.184 |
| 9 | 15.25 | −6.25 | 2.561 |
| | | 0.00 | $\chi^2 = 8.966$ |

Note that the total of the $(O - E)$ column is always zero, thus providing a useful check on the calculation.

If $\chi^2$ exceeds a certain critical value, the null hypothesis is rejected. The critical value depends, as in other significance tests, on the significance level of the test and on the number of degrees of freedom. The number of degrees of freedom is, in an example of this type, one less than the number of classes used, i.e. $4 - 1 = 3$ in this case. The critical values of $\chi^2$ for $P = 0.05$ are given in Table A.5. For 3 degrees of freedom the critical value is 7.815. Since the calculated value of $\chi^2$ is greater than this the null hypothesis is rejected at the 5% significance level: the workers *do* differ in their reliability.

In this calculation of $\chi^2$ it appears that a significant result is obtained because of the high number of breakages reported by the first worker. To study this further, additional chi-squared tests can be performed. One of them tests whether the second, third and fourth workers differ significantly from each other: in this case each expected frequency is $(17 + 11 + 9)/3$. (Note that the $t$-test cannot be used here, as we are dealing with frequencies and not continuous variates). Another tests whether the first worker differs significantly from the other three workers taken as a group. In this case there are two classes: the breakages by the first worker with an expected frequency of 15.25 and the total breakages by the other workers, with expected frequency of $15.25 \times 3 = 45.75$. In such cases when there are only two classes and hence one degree of freedom, an adjustment known as **Yates's correction** should be applied. This involves replacing $O - E$ by $|O - E| - 0.5$, e.g. 4.5 becomes 4. These further tests are given as an exercise at the end of this chapter.

In general the chi-squared test should only be used if the total number of observations is 50 or more and the individual expected frequencies are not less than 5. This is not a rigid rule: a reference is given at the end of this chapter which discusses this point further. Other applications of the chi-squared test are also described in this reference.

### 3.13 TESTING FOR NORMALITY

As has been emphasized in this chapter, many statistical tests assume that the data used are drawn from a normal population. One method of testing this assumption, i.e. using the chi-squared test, was mentioned in the previous section. Unfortunately, this method can only be used if there are 50 or more data points. It is common in experimental work to have only a small set of data. A simple visual way of seeing whether such a set of data is consistent with the assumption of normality is to plot a **cumulative frequency curve** on special graph paper known as **normal probability paper**. This method is most easily explained by means of an example.

*Example.* Use normal probability paper to test whether the values below could have been drawn from a normal population:

109, 89, 99, 99, 107, 111, 86, 74, 115, 107, 134, 113, 110, 88, 104.

Table 3.8 shows the data arranged in order of increasing size. The second column gives the cumulative frequency for each measurement, i.e. the number of measurements less than or equal to that measurement. The third column gives the percentage cumulative frequency. This is calculated by using the formula:

% cumulative frequency = $100 \times$ cumulative frequency$/(n + 1)$

where $n$ is the total number of measurements. (There are mathematical reasons for dividing by $n + 1$ rather than $n$). If the data come from a normal

**Table 3.8**

| Measurement | Cumulative frequency | % cumulative frequency |
|:---:|:---:|:---:|
| 74 | 1 | 6.3 |
| 86 | 2 | 12.5 |
| 88 | 3 | 18.8 |
| 89 | 4 | 25.0 |
| 99 | 5 | 31.3 |
| 99 | 6 | 37.5 |
| 104 | 7 | 43.8 |
| 107 | 8 | 50.0 |
| 107 | 9 | 56.3 |
| 109 | 10 | 62.5 |
| 110 | 11 | 68.8 |
| 111 | 12 | 75.0 |
| 113 | 13 | 81.3 |
| 115 | 14 | 87.5 |
| 124 | 15 | 93.8 |

distribution, a graph of percentage cumulative frequency against measurement results in an S-shaped curve, as shown in Fig. 3.6. Normal probability paper has a non-linear scale for the percentage cumulative frequency axis, which will convert this S-shaped curve into a straight line. The data above,

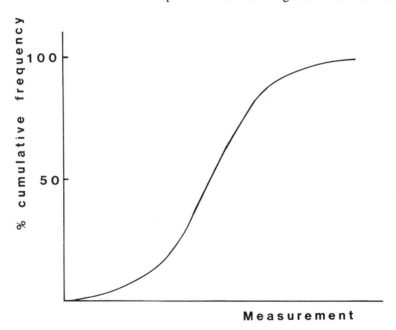

Fig. 3.6 – The cumulative frequency curve for a normal distribution.

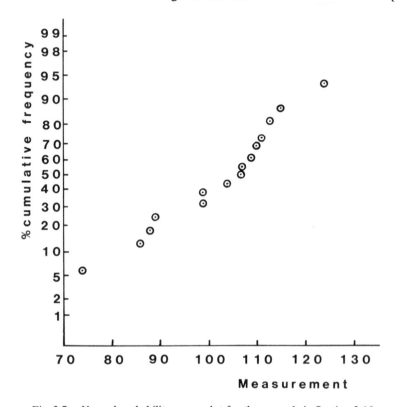

Fig. 3.7 – Normal probability paper plot for the example in Section 3.10.

plotted on such paper, are shown in Fig. 3.7: the points lie approximately on a straight line, supporting the assumption that the data come from a normal distribution.

One method of testing for normality is to use a quantity which measures how closely the points on a normal probability paper plot conform to a straight line. The calculation of this quantity, the product–moment correlation coefficient, $r$, is described in the next chapter (Section 4.3). A reference to the use of $r$ for testing for normality is given at the end of this chapter. This reference also gives a survey of the different tests for normality. Section 5.10 describes another method, the Kolmogorov-Smirnov method, which, among other applications, may be used to test for normality. A worked example is given in that section.

### BIBLIOGRAPHY

V. Barnett and T. Lewis, *Outliers in Statistical Data,* Wiley, New York, 1978. A very comprehensive treatment of the philosophy of outlier rejection and the tests used.

G. E. P. Box, W. G. Hunter and J. S. Hunter, *Statistics for Experimentalists,* Wiley, New York, 1978. Gives further details of testing for significant differences between means as a follow-up to ANOVA.

R. C. Campbell, *Statistics for Biologists,* Cambridge University Press, Cambridge, 1974. Gives tests for homogeneity of variance and normality.

O. L. Davies and P. L. Goldsmith, *Statistical Methods in Research and Production,* Longmans, London, 1972. Gives details of other applications of the chi-squared test and the construction of control charts.

J. J. Filliben, *Technometrics,* 1975, **17**, 111. Describes the use of $r$ in testing for normality, and surveys other tests for normality.

B. R. Kowalski (Ed.), *Chemometrics: Theory and Application,* American Chemical Society, 1977, Washington. Chapter 11 describes tests for normality and discusses the effect of non-normality on parametric tests.

J. C. Miller, *Statistics for Advanced Level,* Cambridge University Press, Cambridge, 1983. Gives an example of the chi-squared test for normality.

R. R. Sokal and F. J. Rohlf, *Biometry,* Freeman, 1969. Gives details of tests for homogeneity of variance.

R. H. Woodward and P. L. Goldsmith, *Cumulative Sum Techniques,* I.C.I. monograph No. 3, Oliver and Boyd, Edinburgh, 1964. A detailed treatment of cusum charts.

## EXERCISES

1. By using a normal probability plot, test whether the data in Table 2.3 could have been drawn from a normal population.

2. The data below are taken from the example in Section 3.3 concerning the concentration of a thiol in blood lysates. Verify that 2.07 is not an outlier.

1.84   1.92   1.94   1.92   1.85   1.91   2.07

3. The following data give the recovery of bromide from spiked samples of vegetable matter, measured by using a gas-liquid chromatographic method. The same amount of potassium bromide was added to each specimen.

Tomato:    777   790   759   790   770   758   764 $\mu g/g$
Cucumber: 782   773   778   765   789   797   782 $\mu g/g$

(J. A. Roughan, P. A. Roughan and J. P. G. Wilkins, *Analyst,* 1983, **108**, 742).

(i) Test whether the recoveries from the two vegetables have variances which differ significantly.
(ii) Test whether the mean recoveries differ significantly.

4. The following results show the percentage of the total available interstitial water recovered by centrifuging samples taken at different depths in sandstone.

| Depth of sample (m) | Water recovered, % |
|---|---|
| 7 | 33.3 33.3 35.7 38.1 31.0 33.3 |
| 8 | 43.6 45.2 47.7 45.4 43.8 46.5 |
| 16 | 73.2 68.7 73.6 70.9 72.5 74.5 |
| 23 | 72.5 70.4 65.2 66.7 77.6 69.8 |

(K. G. Wheatstone and D. Gelsthorpe, *Analyst,* 1982, **107**, 731).
Show that the percentage of water recovered differs significantly at different levels. Use the least significant difference method described in Section 3.9 to find the causes of this significant result.

5. In reading a burette to 0.01 ml the final figure has to be estimated by the analyst. The following frequency table gives the final figures of 50 such readings. Apply an appropriate significance test to determine whether some digits are preferred to others.

    Digit        0 1 2 3 4  5 6 7 8 9
    Frequency 1 6 4 5 3 11 2 8 3 7

6. In order to evaluate a spectrophotometric method for the determination of titanium, the method was applied to alloy samples containing different certified amounts of titanium. The results (% Ti) are shown below.

| Sample | Certified value | Mean spectrophotometric result | Standard deviation |
|---|---|---|---|
| 1 | 0.496 | 0.482 | 0.0257 |
| 2 | 0.995 | 1.009 | 0.0248 |
| 3 | 1.493 | 1.505 | 0.0287 |
| 4 | 1.990 | 2.002 | 0.0212 |

For each alloy 8 replicate determinations were made.
(Qiu Xing-chu and Zhu Ying-quen, *Analyst*, 1983, **108**, 641).
For each alloy, test whether the mean value differs significantly from the certified value.

7. The table below shows further results from the paper cited in Section 3.3 concerning the extraction and determination of tin in foodstuffs (*Analyst*, 1983, **108**, 109). The results give the levels of tin recovered from the same product after boiling for different times in an open vessel.

| Boiling time (min) | Tin found (mg/kg) |
|---|---|
| 30 | 57, 57, 55, 56, 56, 55, 56, 55 |
| 75 | 51, 60, 48, 32, 46, 58, 56, 51 |

Test whether
(a) the variability of the results is greater for the longer boiling time;
(b) the means differ significantly.

8. The data given in the example in Section 3.9 for the number of breakages by four different workers are reproduced below:

$$24, 17, 11, 9$$

Test whether
(a) the number of breakages by the first worker differs significantly from those of the other three workers;
(b) the second, third and fourth workers differ significantly from each other in carefulness.

9. The following results were obtained in a comparison between a new method and the official method of determining oxyphenbutazone. (The figures refer to the percentage recoveries).

|  | New method | Official method |
|---|---|---|
| Mean | 99.35 | 99.53 |
| Variance | 0.185 | 0.152 |
| Sample size | 3 | 3 |

(M. M. Amer, A. H. Taha, B. A. El-Zeany and O. A. El-Sawy, *Analyst*, 1982, **107**, 908).
Test whether the mean results obtained by the two methods differ significantly.

10. The table below gives the amount (in mg/ml) of ephedrine hydrochloride found in pharmaceutical preparations of Ephedrine Elixir B.P., by two different methods: derivative

ultraviolet spectroscopy and an official assay method. (The nominal amount in each sample was 3 mg/ml.)

| Sample | Derivative method | Official method |
|--------|-------------------|-----------------|
| 1 | 2.964 | 2.913 |
| 2 | 3.030 | 3.000 |
| 3 | 2.994 | 3.024 |

(A.G. Davidson and H. Elsheikh, *Analyst*, 1982, **107**, 879).
Test whether the results obtained by the two different methods differ significantly.

11. The following figures refer to the concentration of albumin, in g/l., in the blood sera of 16 healthy adults:

37, 39, 37, 42, 39, 45, 42, 39, 44, 40, 39, 45, 47, 47, 43, 41

(J. W. Foote and H. T. Delves, *Analyst*, 1983, **108**, 492).
The first 8 figures are for men and the second 8 for women. Test whether the mean concentrations for men and women differ significantly.

12. A new flame atomic-absorption spectrometric method of determining antimony in the atmosphere was compared with the recommended colorimetric method. For samples from an urban atmosphere the following results were obtained:

| | Antimony found (mg/m$^3$) | |
|--------|------------|-----------------|
| Sample | New method | Standard method |
| 1 | 22.2 | 25.0 |
| 2 | 19.2 | 19.5 |
| 3 | 15.7 | 16.6 |
| 4 | 20.4 | 21.3 |
| 5 | 19.6 | 20.7 |
| 6 | 15.7 | 16.8 |

(J. R. Catillo, J. Lanaja, M$^a$. C. Martinez and J. Aznárez, *Analyst*, 1982, **107**, 1488).
Do the results obtained by the two methods differ significantly?

13. The data in the table below give the concentrations of albumin measured in the blood serum of one adult. On each of four consecutive days a blood sample was taken and three replicate determinations of the serum albumin concentration were made.

| Day | Albumin concentration (normalized, arbitrary units) | | |
|-----|----|----|----|
| 1 | 63 | 61 | 62 |
| 2 | 57 | 56 | 56 |
| 3 | 50 | 46 | 46 |
| 4 | 57 | 54 | 59 |

Show that the mean concentrations for different days differ significantly. Estimate the variance of the day-to-day variation (i.e. the 'sampling variation').

# 4

# Errors in instrumental analysis; regression and correlation

## 4.1 INSTRUMENTAL ANALYSIS

Classical or "wet chemistry" analysis techniques such as titrimetry and gravimetry remain in use in many laboratories and are still widely taught in Analytical Chemistry courses. They provide excellent introductions to the manipulative and other skills required in analytical work; they are ideal for high-precision analyses, especially when small numbers of samples are involved; and they are necessary for the analysis of standard materials. However, there is no doubt that most analyses are now performed by instrumental methods. Techniques using absorption and emission spectroscopy at various wavelengths, several different electrochemical methods, mass spectrometry, gas and liquid chromatography, and thermal and radiochemical methods, probably account for as much as 90% of all current analytical work, or even more. There are several reasons for this.

(1) Instrumental methods can perform analyses which are difficult or impossible by classical methods. Whereas classical methods can only rarely detect materials at sub-microgram levels, many instrumental methods are astonishingly sensitive. For example luminescence methods have detected organic molecules at the $10^{-18}M$ level. It is normally only possible to determine one analyte at a time by "wet chemical" methods, but plasma spectrometry can determine ten or more elements simultaneously (and at very low concentrations), while methods which combine high-performance liquid chromatography with a spectroscopic detection procedure can identify and determine many components of complex organic mixtures within a few minutes. Furthermore, the concentration range of a particular classical analysis method is usually limited by practical and theoretical considerations. Thus EDTA titrations can be successfully performed with reactant concentrations as low as about $10^{-4}M$, but an upper limit is set by the solubility of EDTA in water. The useful concentration range is generally 2–3 orders of magnitude (i.e. powers of ten) for classical methods. In contrast, *some* instrumental methods are able to determine analyte concentrations over a range of six or more orders of magnitude: this characteristic has important

implications for the statistical treatment of the results, as we shall see in the next section.

(2) For a large through-put of samples, instrumental analysis is generally quicker and often cheaper than the labour-intensive manual methods. In clinical analysis, for example, there is frequently a requirement for the same analyses to be done on scores or even hundreds of blood or blood serum/plasma samples every day. Despite the high initial cost of the equipment, such work is generally performed by using completely automatic systems. Automation has become such an important feature of analytical chemistry that the ease with which a particular technique can be automated often determines whether or not it is used at all. A typical automatic method may be able to process samples at the rate of 100 per hour or more: the equipment will take a measured volume of a sample, dilute it appropriately, conduct one or more reactions on it, and determine and record the concentration of the analyte or a derivative produced in the reactions. Special problems of error estimation will evidently arise in automatic analysis; systematic errors, for example, must be identified and corrected as rapidly as possible.

(3) Modern analytical instruments are readily interfaced with computers. The present tendency is to use 8-bit or 16-bit microcomputers with floppy-disc or hard-disc storage devices. Such systems can provide a reasonably sophisticated degree of instrument control and data-handling. The computer can, for example, control the timing of sampling or injection apparatus by activating electrical switches. When the analytical results have been obtained the computer can process the data, performing such operations as the generation of derivative spectra or Fourier transforms. It can also evaluate the results statistically, the recurring theme of this book, and can compare the analytical results with data stored in its memory so as to match spectral and other data. All these facilities are available from low-cost computers operating at acceptably high speeds. A further important possibility is the development of 'intelligent' instruments which, for example, perform automatic optimization processes (see Chapter 6). There is no doubt that computer facilities will soon be provided with virtually all analytical instruments, further enhancing their capabilities and efficiency.

## 4.2 CALIBRATION GRAPHS IN INSTRUMENTAL ANALYSIS

The ability of instrumental analysis techniques to handle a wide range of analyte concentrations means that the results are calculated, and the random errors evaluated, in a particular fashion that differs from that used when a single measurement is repeated several times. The usual procedure is as follows. The analyst takes a series of samples (normally at least three or four, and possibly several more) in which the concentration of the analyte is *known*. These calibration standards are measured in the analytical instrument under the same conditions as those subsequently used for the test (i.e. the 'unknown') samples. Once the calibration graph has been established the analyte concentration in any test sample can be obtained, as shown in Fig. 4.1, by interpolation. This general procedure raises several important statistical questions.

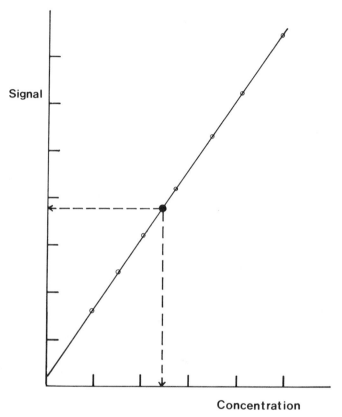

Fig. 4.1 – Calibration procedure in instrumental analysis; ○ calibration points; ● test sample.

(1) Is the calibration graph linear? If it is a curve, what is the form of the curve?

(2) Bearing in mind that each of the points on the calibration graph is subject to errors, what is the best straight line (or curve) through these points?

(3) Assuming that the calibration plot is actually linear, what are the estimated errors and confidence limits for the slope and the intercept of the line?

(4) When the calibration plot is used for the analysis of a test sample, what are the errors and confidence limits for the determined concentration?

(5) What is the **limit of detection** of the method? That is, what is the least concentration of the analyte that can be detected with a predetermined level of confidence?

Before tackling these questions in detail, we must consider a number of important aspects of plotting calibration graphs. First, it is usually essential that the calibration standards cover the whole range of concentrations required in the subsequent analyses. With the important exception of the 'method of standard additions', which is treated separately in a later section, concentrations of test samples are normally determined by interpolation and *not* by extrapolation. Secondly it is crucially important to include the value for a 'blank' sample in the calibration curve. The blank contains no deliberately added analyte, but does contain the same solvents, reagents etc. as the other test samples, and is subjected to exactly the same sequence of analytical procedures. The instrument signal given by the blank sample will often not be zero. It is, of course, subject to errors, like all the other points on the calibration plot, and it is thus wrong in principle to subtract the blank value from the other standard values before plotting the calibration graph. Finally, it should be noted that the calibration curve is always plotted with the instrument response on the vertical ($y$) axis and the standard concentrations on the horizontal ($x$) axis. This is because many of the procedures to be described in the following sections *assume* that all the errors are in the $y$-values and that the standard concentrations ($x$-values) are error-free. This assumption is further discussed below.

The reader should note that the methods now to be described continue to generate a good deal of controversy, evidenced by many scientific meetings and original papers on the subject of calibration. Much of this controversy derives from the fact that the generally used methods rest on two assumptions. The first is that mentioned in the previous paragraph, viz. that the errors in the calibration experiment occur only in the $y$-values. In many routine instrumental analyses this assumption may well be justified. The standards can be made up with an error of ca. 0.1% or better (see Chapter 1), whereas the instrumental measurements themselves might have a coefficient of variation of 1-2% or worse. In recent years the advent of some high-precision automatic instruments has again put the assumption under question, however, and has led some users to make up their standard solutions by weight rather than by the less accurate combination of weight and volume.

The second assumption is that the magnitude of the errors in $y$ is independent of the analyte concentration. Common sense indicates that this is often unlikely to be true: if, for example, the relative errors in measurement are constant, the absolute errors will increase as the analyte concentration increases. As we shall see, it is possible to modify the statistical procedure to take such factors into account; in practice, however, the modified (and rather more complex) procedures seem not to be used in many cases where they should be adopted.

## 4.3 THE PRODUCT-MOMENT CORRELATION COEFFICIENT

In this section we discuss the first problem listed in the previous section — is the calibration plot linear? We assume that a straight-line plot takes the algebraic form:

$$y = bx + a \qquad (4.1)$$

where $b$ is the slope of the line and $a$ its intercept on the $y$-axis. The individual points on the line will be referred to as $(x_1, y_1)$ (normally the "blank" reading), $(x_2, y_2), (x_3, y_3) \ldots (x_i, y_i) \ldots (x_n, y_n)$, i.e. there are $n$ points altogether. The mean of the $x$-values is, as usual, called $\bar{x}$, and the mean of the $y$-values is $\bar{y}$: the position $(\bar{x}, \bar{y})$ is then known as the "centroid" of all the points.

In order to estimate how well the experimental points fit a straight line, we calculate the **product–moment correlation coefficient**, $r$. This statistic is often referred to simply as the 'correlation coefficient' because in quantitative sciences it is by far the most commonly used type of correlation coefficient. We shall, however, meet other types of correlation coefficient in Chapter 5. The value of $r$ is given by:

$$r = \frac{\sum_i \left\{ (x_i - \bar{x})(y_i - \bar{y}) \right\}}{\left\{ [\sum_i (x_i - \bar{x})^2] [\sum_i (y_i - \bar{y})^2] \right\}^{\frac{1}{2}}} \qquad (4.2)$$

Close study of this equation shows that $r$ can take values in the range $-1 \leqslant r \leqslant +1$. As indicated in Fig. 4.2, an $r$ value of $-1$ describes perfect negative correlation, i.e. all the experimental points lie on a straight line of negative slope. Similarly, when $r = +1$ we have perfect positive correlation, all the points lying exactly on a straight line of positive slope. When there is no correlation between $x$ and $y$, the value of $r$ is zero. In analytical practice, calibration graphs frequently give numerical $r$ values of greater than 0.99, and $r$ values of less than about 0.90 are relatively uncommon. A typical example of a calculation of $r$ illustrates a number of important points.

*Example.* Standard aqueous solutions of fluorescein' are examined in a fluorescence spectrometer, and yield the following fluorescence intensities (in arbitrary units):

| Fluorescence intensities: | 2.1 | 5.0 | 9.0 | 12.6 | 17.3 | 21.0 | 24.7 |
|---|---|---|---|---|---|---|---|
| Concentration, pg/ml: | 0 | 2 | 4 | 6 | 8 | 10 | 12 |

Determine the correlation coefficient, $r$.

In practice, such calculations will almost certainly be performed on a programmable calculator or microcomputer, but it is important and instructive to examine a manually calculated result. The data are presented in a table, as follows.

| $x_i$ | $y_i$ | $x_i - \bar{x}$ | $(x_i - \bar{x})^2$ | $y_i - \bar{y}$ | $(y_i - \bar{y})^2$ | $(x_i - \bar{x})(y_i - \bar{y})$ |
|---|---|---|---|---|---|---|
| 0 | 2.1 | −6 | 36 | −11.0 | 121.00 | 66.0 |
| 2 | 5.0 | −4 | 16 | −8.1 | 65.61 | 32.4 |
| 4 | 9.0 | −2 | 4 | −4.1 | 16.81 | 8.2 |
| 6 | 12.6 | 0 | 0 | −0.5 | 0.25 | 0 |
| 8 | 17.3 | 2 | 4 | 4.2 | 17.64 | 8.4 |
| 10 | 21.0 | 4 | 16 | 7.9 | 62.41 | 31.6 |
| 12 | 24.7 | 6 | 36 | 11.6 | 134.56 | 69.6 |
| 42 | 91.7 | 0 | 112 | 0 | 418.28 | 216.2 |

$$\bar{x} = 42/7 = 6; \quad \bar{y} = 91.7/7 = 13.1$$

The figures below the line at the foot of the columns are in each case the sums of the figures in the table. [Note that $\sum_i (x_i - \bar{x})$ and $\sum_i (y_i - \bar{y})$ are both zero.] Using these totals in conjunction with Eq. (4.2), we have

$$
\begin{aligned}
r &= 216.2/(112 \times 418.28)^{\frac{1}{2}} \\
&= 216.2/216.44 \\
&= 0.9989
\end{aligned}
$$

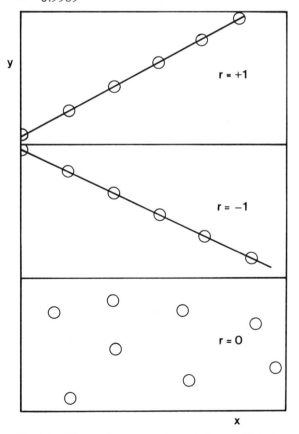

Fig. 4.2 – The product–moment correlation coefficient, $r$.

Two observations follow from this example. As is shown in Fig. 4.3, although several of the points deviate noticeably from the 'best' straight line (calculated by using the principles of the following section), the *r* value is very close indeed to 1. Experience shows that even quite poor-looking calibration plots give very high *r* values. In such cases the numerator and denominator in Eq. (4.2) are nearly equal. It is thus very important to perform the calculation with an adequate number of significant figures. In the example given, neglect of the figures after

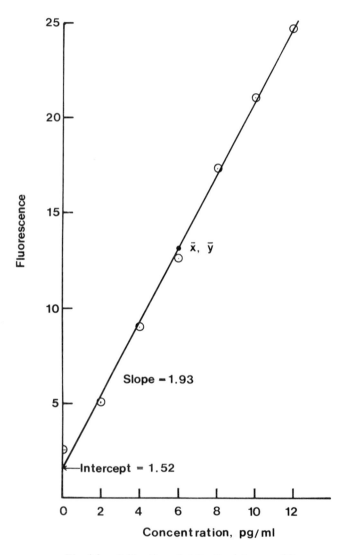

Fig. 4.3 – Calibration plot for the data on p. 89.

the decimal point would have given an obviously incorrect $r$ value of exactly 1, and the use of only one place of decimals would have given the incorrect $r$ value of 0.9991. This point is especially important when a calculator or computer is used to determine $r$; such devices do not always provide sufficient figures.

Although correlation coefficients are simple to calculate, they are all too easily misinterpreted. It must always be borne in mind that the use of Eq. (4.2) will generate an $r$ value even if the data are patently non-linear in character. Figure 4.4 shows two examples in which a calculation of $r$ would be misleading. In Fig. 4.4a, the points of the calibration plot clearly lie on a curve; this curve is sufficiently gentle, however, to yield quite a high correlation coefficient when Eq. (4.2) is applied. The lesson of this example is that the calibration curve must *always* be physically plotted (on graph paper or a computer monitor): otherwise

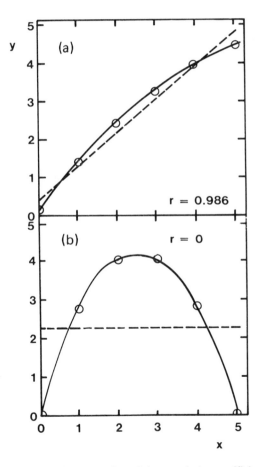

Fig. 4.4 – Misinterpretation of the correlation coefficient, $r$.

a straight-line relationship might wrongly be deduced from the calculation of $r$. Figure 4.4b is a reminder that a zero correlation coefficient does not mean that $y$ and $x$ are entirely unrelated; it only means that they are not *linearly* related.

As we have seen, $r$ values obtained in instrumental analysis are normally very high, so a calculated value, together with the calibration plot itself, is often sufficient to assure the analyst that he has indeed obtained a useful linear relationship. In some circumstances, however, much lower $r$ values are obtained; one such situation is further discussed in Section 4.9. In these cases it will be necessary to use a proper statistical test to see whether the correlation coefficient is indeed significant, bearing in mind the number of pairs of points used in the calculation. The simplest method of doing this is to calculate a $t$-value (see Chapter 3 for a fuller discussion of the $t$-test), using the equation

$$t = \frac{|r|\sqrt{(n-2)}}{\sqrt{(1-r^2)}} \tag{4.3}$$

The calculated value of $t$ is compared with the tabulated value at the desired significance level, using a *two-tailed* $t$-test and $(n-2)$ degrees of freedom. The null hypothesis in this case is that there is no correlation between $x$ and $y$. If the calculated value of $t$ is greater than the tabulated value, the null hypothesis is rejected; that is, we conclude in such a case that a significant correlation does exist.

## 4.4 THE LINE OF REGRESSION OF $y$ ON $x$

In this section we assume that there is a linear relationship between the analytical signal $(y)$ and the concentration $(x)$, and show how to calculate the 'best' straight line through the calibration graph points, each of which is subject to experimental error. Since we are assuming for the present that all the errors are in $y$ (cf. Section 4.2 above), we are seeking the line that minimizes the deviations in the $y$-direction between the experimental points and the calculated line. Since some of these deviations (technically known as the $y$-residuals) will be positive and some negative, it is sensible to seek to minimize the **sum of the squares of the residuals**. This explains the frequent use of the term 'method of least squares' for the procedure. The straight line required is calculated on this principle: as a result it is found that the line must pass through the "centroid" of the points, $(\bar{x}, \bar{y})$. It can be shown that

$$b = \frac{\sum_i \left\{ (x_i - \bar{x})(y_i - \bar{y}) \right\}}{\sum_i (x_i - \bar{x})^2} \tag{4.4}$$

$$a = \bar{y} - b\bar{x} \tag{4.5}$$

The line thus calculated is known as the **line of regression of** $y$ **on** $x$, i.e. the line indicating how $y$ varies when $x$ is set to chosen values. It is very important to notice that the line of regression of $x$ on $y$ *is not the same line* (except in the highly improbable case where all the points lie exactly on a straight line, when $r = 1$ exactly). The line of regression of $x$ on $y$ (which also passes through the centroid of the points) assumes that all the errors occur in the $x$-direction. If we maintain rigidly the convention that the analytical signal is always plotted on the $y$-axis and the concentration on the $x$-axis, it is always the line of regression of $y$ on $x$ that we must use in calibration experiments.

*Example.* Calculate the slope and intercept of the regression line for the data given in the previous example (see Section 4.3).

In Section 4.3 we calculated that, for this calibration curve:

$$\sum_i (x_i - \bar{x})(y_i - \bar{y}) = 216.2; \quad \sum_i (x_i - \bar{x})^2 = 112; \quad \bar{x} = 6; \quad \bar{y} = 13.1.$$

Using Eqs. (4.4) and (4.5) we calculate that

$$b = 216.2/112 = 1.93$$

$$a = 13.1 - (1.93 \times 6) = 13.1 - 11.58 = 1.52$$

The equation for the regression line is $y = 1.93x + 1.52$.

It is apparent that the terms used to calculate $r$ are also needed in the calculation of $a$ and $b$: this facilitates the calculation of all these terms with a calculator or simple computer program.

The results of the slope and intercept calculations are depicted in Fig. 4.3. Again it is important to emphasize that Eqs. (4.4) and (4.5) must not be misused — they will only give useful results when prior study (calculation of $r$ and a visual inspection of the points) has indicated that a straight line relationship is realistic for the experiment in question. Non-parametric methods (i.e. methods that make no assumptions about the nature of the error distribution) can also be used to calculate regression lines, and this topic is treated in the next chapter.

## 4.5 ERRORS IN THE SLOPE AND INTERCEPT OF THE REGRESSION LINE

The line of regression calculated in the previous section will in practice be used to estimate the concentrations of test samples by interpolation, and perhaps also to estimate the limit of detection of the analytical procedure. The random errors in the values for the slope and intercept are thus of importance, and the equations used to calculate them are now considered. We must first calculate the statistic $s_{y/x}$, which is given by:

$$s_{y/x} = \left\{ \frac{\sum_i (y_i - \hat{y})^2}{n - 2} \right\}^{\frac{1}{2}} \tag{4.6}$$

It will be seen that this equation utilizes the $y$-residuals, $y_i - \hat{y}$, where the $\hat{y}$ values are the points on the calculated regression line corresponding to the individual $x$-values, i.e. the 'fitted' $y$-values (Fig. 4.5). The $\hat{y}$-value for a given value of $x$ is of course readily calculated from the regression equation. Equation (4.6) is clearly similar in form to the equation for the standard deviation of a set of repeated measurements [Eq. (2.2)]; the former differs in that deviations, $(y_i - \bar{y})$, are replaced by residuals, $(y_i - \hat{y})$, and the denominator contains the term $(n - 2)$ rather than $(n - 1)$. In a regression calculation the number of degrees of freedom (cf. Section 2.4) is $(n - 2)$. This clearly reflects the obvious consideration that only one straight line can be drawn through two points.

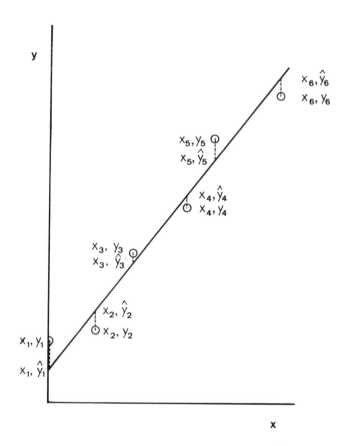

Fig. 4.5 – The $y$-residuals of a regression line.

Armed with a value for $s_{y/x}$, we can now calculate $s_b$ and $s_a$, the standard deviations for the slope ($b$) and the intercept ($a$). These are given by:

$$s_b = \frac{s_{y/x}}{\left\{\sum_i (x_i - \bar{x})^2\right\}^{\frac{1}{2}}} \qquad (4.7)$$

$$s_a = s_{y/x}\left\{\frac{\sum_i x_i^2}{n \sum_i (x_i - \bar{x})^2}\right\}^{\frac{1}{2}} \qquad (4.8)$$

The values of $s_b$ and $s_a$ can be used in the usual way (cf. Chapter 2) to estimate confidence limits for the slope and intercept. Thus the confidence limits for the slope are given by $b \pm ts_b$, where the $t$-value is taken at the desired confidence level and $(n - 2)$ degrees of freedom. Similarly the confidence limits for the intercept are given by $a \pm ts_a$.

*Example.* Calculate the standard deviations and confidence limits of the slope and intercept of the regression line calculated in Section 4.4.

This type of calculation may not be directly accessible on a programmable calculator, although useful computer programs have been developed (see the bibliography for Chapter 1). Here we perform the calculation manually, using a tabular lay-out as before.

| $x_i$ | $x_i^2$ | $y_i$ | $\hat{y}$ | $|y_i - \hat{y}|$ | $(y_i - \hat{y})^2$ |
|---|---|---|---|---|---|
| 0 | 0 | 2.1 | 1.52 | 0.58 | 0.3364 |
| 2 | 4 | 5.0 | 5.38 | 0.38 | 0.1444 |
| 4 | 16 | 9.0 | 9.24 | 0.24 | 0.0576 |
| 6 | 36 | 12.6 | 13.10 | 0.50 | 0.2500 |
| 8 | 64 | 17.3 | 16.96 | 0.34 | 0.1156 |
| 10 | 100 | 21.0 | 20.82 | 0.18 | 0.0324 |
| 12 | 144 | 24.7 | 24.68 | 0.02 | 0.0004 |

$$\sum_i x_i^2 = 364 \qquad\qquad \sum_i (y_i - \hat{y})^2 = 0.9368$$

From the table and using Eq. (4.6) we obtain

$$s_{y/x} = \sqrt{0.9368/5} = \sqrt{0.18736} = 0.4329.$$

From Section 4.3 we know that $\sum_i (x_i - \bar{x})^2 = 112$, and Eq. (4.7) can be used to show that

$$s_b = 0.4329/\sqrt{112} = 0.4329/10.58 = 0.0409.$$

The $t$-value for $(n - 2) = 5$ and the 95% confidence level is 2.57 (Table A.1). The 95% confidence limits for $b$ are thus

$$b = 1.93 \pm 2.57 \times 0.0409 = 1.93 \pm 0.11$$

Equation (4.8) requires knowledge of $\sum_i x_i^2$, calculated as 364 from the table. We can thus write

$$s_a = 0.4329\sqrt{(364/784)} = 0.2950$$

so the confidence limits are

$$a = 1.52 \pm 2.57 \times 0.2950 = 1.52 \pm 0.76$$

In this example, the number of significant figures necessary was not large, but it is always a useful precaution to use the maximum available number of significant figures during such a calculation, rounding only at the end.

## 4.6 CALCULATION OF A CONCENTRATION

Once the slope and intercept of the regression line have been determined, it is very simple to calculate an $x$-value corresponding to any measured $y$-value. A more complex problem arises when it is necessary to estimate the error in a concentration calculated by using a regression line. Calculation of an $x$-value from a given $y$-value involves the use of both the slope ($b$) and the intercept ($a$) and, as we saw in the previous section, both these values are subject to error. As a result, the determination of the error in the $x$-value is extremely complex, and most workers use the following approximate formula:

$$s_{x_0} = \frac{s_{y/x}}{b} \left\{ 1 + \frac{1}{n} + \frac{(y_0 - \bar{y})^2}{b^2 \sum_i (x_i - \bar{x})^2} \right\}^{\frac{1}{2}} \tag{4.9}$$

In this equation, $y_0$ is the experimental value of $y$ from which the concentration value $x_0$ is to be determined, $s_{x_0}$ is the estimated standard deviation of $x_0$, and the other symbols have their usual meaning. In some cases an analyst may make several readings to obtain the value of $y_0$; if there are $m$ such readings, then the equation for $s_{x_0}$ becomes

$$s_{x_0} = \frac{s_{y/x}}{b} \left\{ \frac{1}{m} + \frac{1}{n} + \frac{(y_0 - \bar{y})^2}{b^2 \sum_i (x_i - \bar{x})^2} \right\}^{\frac{1}{2}} \tag{4.10}$$

As expected, Eq. (4.10) reduces to Eq. (4.9) if $m = 1$. As always, confidence limits can be calculated as $x_0 \pm ts_{x_0}$, $(n-2)$ degrees of freedom. Again, a simple computer program will perform all these calculations, but most preprogrammed calculators will not be wholly adequate.

*Example.* Using the data from the example in Section 4.3, determine the $x_0$ and $s_{x_0}$ values and $x_0$ confidence limits for solutions with fluorescence intensities 2.9, 13.5, and 23.0 units.

The $x_0$ values are easily calculated by using the regression equation determined in Section 4.4, $y = 1.93x + 1.52$. Substituting the $y_0$-values 2.9, 13.5 and 23.0, we obtain $x_0$-values of 0.72, 6.21 and 11.13 pg/ml respectively.

To obtain the $s_{x_0}$-values corresponding to these $x_0$-values we use Eq. (4.9), recalling from the preceding sections that $n = 7$, $b = 1.93$, $s_{y/x} = 0.4329$, $\bar{y} = 13.1$, and $\sum_i (x_i - \bar{x})^2 = 112$. The $y_0$ values 2.9, 13.5 and 23.0 then yield $s_{x_0}$-values of 0.26, 0.24 and 0.26 respectively. The corresponding 95% confidence limits ($t = 2.57$) are $0.72 \pm 0.68$, $6.21 \pm 0.62$, and $11.13 \pm 0.68$ pg/ml respectively.

This example illustrates a point of some importance. It is apparent that the confidence limits are rather smaller (i.e. better) for the result $y_0 = 13.5$ than for the other two $y_0$-values. Inspection of Eq. (4.9) confirms that as $y_0$ approaches $\bar{y}$, the third term inside the bracket approaches zero, and $s_{x_0}$ thus approaches a minimum value. The general form of the confidence limits for a calculated concentration is shown in Fig. 4.6. In a practical analysis, therefore, a calibration

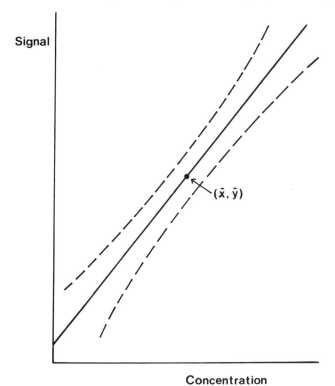

**Concentration**

Fig. 4.6 – General form of the confidence limits for a concentration determined by using an unweighted regression line.

experiment of this type will give the most precise results when the measured instrument signal corresponds to a point close to the centroid of the regression line.

If we wished to improve (i.e. narrow) the confidence limits in this calibration experiment, Eqs. (4.9) and (4.10) show that at least two approaches should be considered. We could increase $n$, the number of calibration points on the regression line, and we could make more than one measurement of $y_0$, and use the mean value of $m$ such measurements in the calculation of $x_0$. The results of such procedures can be assessed by considering the three terms inside the brackets in the two equations. In the example above, the dominant term in all three calculations is the first one – unity. It follows that in this case (and many others) an improvement in precision might be made by measuring $y_0$ several times and using Eq. (4.10) rather than Eq. (4.9). If, for example, the $y_0$-value of 13.5 had been calculated as the mean of four determinations, then the $s_{x_0}$-value and the confidence limits would have been 0.14 and 6.21 ± 0.36 respectively, both results indicating substantially improved precision. Of course, making too many replicate measurements (assuming that sufficient sample is available) generates much more work for only a small additional benefit: the reader should verify that 8 measurements of $y_0$ would produce an $s_{x_0}$-value of 0.12 and confidence limits of 6.21 ± 0.30.

The effect of $n$, the number of calibration points, on the confidence limits of the concentration determination is more complex. This is because we also have to take into account accompanying changes in the value of $t$. Use of a large number of calibration samples involves the task of preparing many accurate standards for only marginally increased precision (cf. the effects of increasing $m$, described in the previous paragraph). On the other hand, small values of $n$ are not permissible: in such cases, not only will $1/n$ be larger, but the number of degrees of freedom, $(n - 2)$, will become very small, necessitating the use of very large $t$ values in the calculation of the confidence limits. In many experiments, as in the example given, six or so calibration points will be adequate, the analyst gaining extra precision, if necessary, by repeated measurements of $y_0$.

## 4.7 LIMITS OF DETECTION

As we have seen, one of the principal benefits of using instrumental methods of analysis is that they are capable of detecting and determining much smaller quantities of analyte than classical analysis methods. These benefits have led to the appreciation of the importance of trace concentrations of many materials, for example in biological and environmental samples, and thus to the development of many further techniques in which low limits of detection are a major criterion of successful application. It is therefore evident that statistical methods for assessing and comparing limits of detection are of importance. In general terms, the limit of detection of an analyte may be described as that concentration which gives an instrument signal ($y$) *significantly different* from the 'blank' or 'background' signal. It is immediately apparent that this description gives the analyst a good deal of freedom to decide the exact definition of the limit of detection, based on a suitable interpretation of the phrase 'significantly different'. There has in practice been very little agreement between professional and statutory bodies on this point, and the whole area is as controversial as other aspects of the statistical treatment of calibration analyses. A commonly used definition in the literature of analytical chemistry is that the limit of detection is *the analyte concentration giving a signal equal to the blank signal, $y_B$, plus two standard deviations of the blank, $s_B$.* Recent guidelines from (particularly American) public bodies suggests that the criterion should be:

$$y - y_B = 3s_B \qquad (4.11)$$

The significance of this last definition is illustrated in more detail in Fig. 4.7. An analyst studying trace concentrations is confronted with two problems: he does not wish to claim the presence of the analyte when it is actually absent, but equally he does not wish to report that the analyte is absent when it is in fact present. The possibility of each of these errors must be minimized under a sensible definition of a limit of detection. In the figure, curve A represents the normal distribution of measured values of the blank signal. It would be possible to identify a point, $y = P$, towards the upper edge of this distribution, and claim that a signal greater than this was unlikely to be due to the blank (Fig. 4.7),

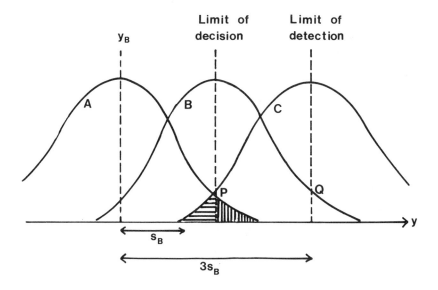

Fig. 4.7 — Definitions of (a) the limit of decision; (b) the limit of detection.

whereas a signal less than P would be assumed to indicate a blank sample. However, for a sample giving an average signal P, 50% of the observed signals will be less than this, since the signal will have a normal distribution (of the same shape as that for the blank — see below) extending below P (curve B). The probability of concluding that this sample does not differ from the blank when in fact it *does,* is therefore 50%. Point P, which has been called the limit of decision, is thus unsatisfactory as a limit of detection, since it solves the first of the problems mentioned above, but not the second. A more suitable point is at $y = Q$ (Fig. 4.7), such that Q is twice as far as P from $y_B$. It may be shown that if $y_B$-Q is 3.28 times the standard deviation of the blank, $s_B$, then the probability of each of the two kinds of error occurring (indicated by the shaded areas in Fig. 4.7) is only 5%. If, as suggested by Eq. (4.11), the distance $y_B$-Q is only $3s_B$, the probability of each error is about 7%; many analysts would consider that this is a reasonable definition of a limit of detection.

It must be re-emphasized that this definition of a limit of detection is quite arbitrary, and it is entirely open to an analyst to provide an alternative definition for a particular purpose. For example, there may be occasions when an analyst is anxious to avoid at all costs the possibility of reporting the absence of the analyte when it is in fact present, but is relatively unworried about the opposite error. It is clear that, whenever a limit of detection is cited in a paper or report, the definition used to obtain it must also be provided. Some attempts have been made to define a further limit, the 'limit of quantitation' (or 'limit of determination'), which is regarded as the lower limit for precise quantitative measurements, as opposed to qualitative detection. A value of $y_B + 10s_B$ has been suggested for this limit, but it has not been very widely used in practice.

We must now discuss how the terms $y_B$ and $s_B$ are obtained in practice when a conventional regression line is used for calibration as described in the preceding sections. A fundamental assumption of the unweighted least-squares method that we have studied thus far in this chapter is that each point on the plot (including the point representing the blank or background) has a normally distributed variation (in the $y$-direction only) with a standard deviation estimated by $s_{y/x}$ [Eq. (4.6)]. This is the justification for drawing the normal distribution curves with the same width in Fig. 4.7. It is therefore appropriate to use $s_{y/x}$ in place of $s_B$ in the estimation of the limit of detection. It is, of course, possible to perform the blank experiment several times and obtain an independent value for $s_B$, but this is a time-consuming procedure, and the use of $s_{y/x}$ is quite suitable in practice. The value of $a$, the calculated intercept, can be used as an estimate of $y_B$, the blank signal itself; it should be a more accurate estimate of $y_B$ than the single measured blank value, $y_1$.

*Example.* Estimate the limit of detection for the fluorescein determination studied in the previous sections.

We use Eq. (4.11) with the values of $y_B$ (= $a$) and $s_B$ (= $s_{y/x}$) previously calculated. The value of $y$ at the limit of detection is found to be $1.52 + 3 \times 0.4329$, i.e. 2.82. Use of the regression equation then yields a detection limit of 0.67 pg/ml. Figure 4.8 summarizes all the calculations performed on the fluorescein determination data.

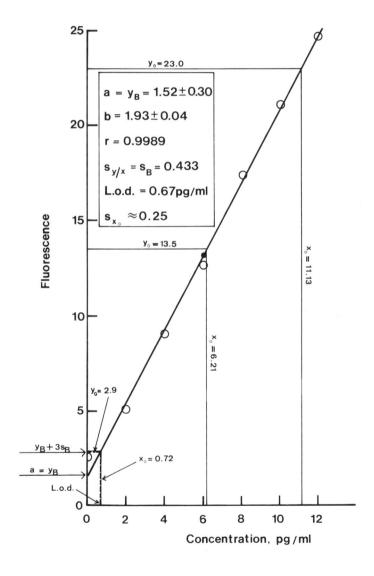

Fig. 4.8 – Summary of the calculations using the data on p. 87.

It is very important to avoid confusing the limit of detection of a technique with its **sensitivity**. This very common source of confusion probably arises because there is no generally-accepted English word synonymous with 'having a low limit of detection'. The word 'sensitive' is generally used for this purpose, thus giving rise to much ambiguity. The sensitivity of a technique is correctly defined as the *slope* of the calibration graph and, provided the plot is linear, can be measured at any point on it. In contrast, the limit of detection of a method is calculated with the aid of the section of the plot close to the origin, and utilizes both the slope and the intercept.

## 4.8 THE METHOD OF STANDARD ADDITIONS

Suppose that an analyst wishes to determine the concentration of silver in samples of photographic waste by atomic-absorption spectrometry. Using the methods of the previous sections, he could calibrate the spectrometer with some aqueous solutions of a pure silver salt and use the resulting calibration graph in the determination of the silver in the test samples. This method is only valid, however, if a pure aqueous solution of silver gives the same absorption signal as a photographic waste sample containing the same concentration of silver. In other words, in using pure solutions to establish the calibration graph it is assumed that there are no 'matrix effects', i.e. reduction or enhancement of the silver absorbance signal by other sample components. In many areas of analysis, such an assumption is frequently invalid. Matrix effects occur even with methods such as plasma spectrometry that have a reputation for being relatively free from interferences.

The first possible solution to this problem is to take a sample of photographic waste that is similar to the test sample but free from silver, and add known amounts of a silver salt to it to make up the standard solutions. The calibration graph will then be set up by using an apparently suitable matrix. In many cases, however, this approach is impracticable. It will not eliminate matrix effects that differ in magnitude from one sample to another, and it may not be possible to obtain a sample of the matrix that contains no analyte – for example, a silver-free sample of photographic waste is unlikely to occur! It follows that all the analytical measurements, including the establishment of the calibration graph, must in some way be performed with the sample itself. This is achieved in practice by using the **method of standard additions**. The method is widely practised in atomic absorption and emission spectrometry and has also found application in electrochemical analysis and other areas. Equal volumes of the sample solution are taken, all but one are separately 'spiked' with known and different amounts of the analyte, and *all* are then diluted to the same volume. The instrument signals are then determined for all these solutions and the results plotted as shown in Fig. 4.9. As usual, the signal is plotted on the $y$-axis; in this case the $x$-axis is graduated in terms of the amounts of analyte *added* (either as an absolute weight or as a concentration). The (unweighted) regression line is

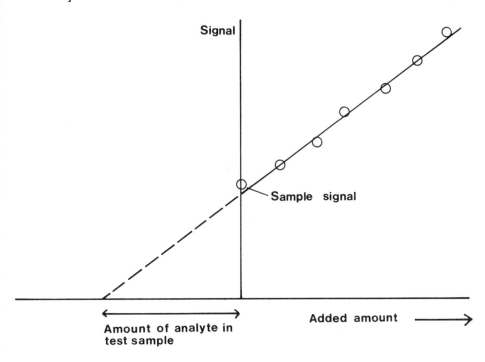

Fig. 4.9 – The method of standard additions.

calculated in the normal way, but space is provided for it to be extrapolated back to the point on the $x$-axis at which $y = 0$. It is clear that this negative intercept on the $x$-axis corresponds to the amount of the analyte in the test sample. Inspection of the figure shows that this value is given by $a/b$, the ratio of the intercept and the slope of the regression line. Since both $a$ and $b$ are subject to error (Section 4.5) the calculated value is clearly subject to error as well. In this case, however, the amount is not predicted from a single measured value of $y$, so the formula for the standard deviation, $s_{x_E}$, of the extrapolated $x$-value ($x_E$) is not the same as that in Eq. (4.9). Instead, we use:

$$s_{x_E} = \frac{s_{y/x}}{b} \left\{ \frac{1}{n} + \frac{\bar{y}^2}{b^2 \sum_i (x_i - \bar{x})^2} \right\}^{\frac{1}{2}} \tag{4.12}$$

Increasing the value of $n$ again improves the precision of the estimated amount: in general, at least six points should be used in a standard-additions experiment. Moreover, the precision is improved by maximizing $\sum_i (x_i - \bar{x})^2$, so the calibration solutions should, if possible, cover a considerable range. Confidence limits for $x_E$ can, as always, be determined as $x_E \pm t s_{x_E}$.

*Example.* The silver concentration in a sample of photographic waste was determined by atomic-absorption spectrometry with the method of standard additions. The following results were obtained.

| Added Ag, $\mu$g per ml of original sample solution | 0 | 5 | 10 | 15 | 20 | 25 | 30 |
|---|---|---|---|---|---|---|---|
| Absorbance | | 0.32 | 0.41 | 0.52 | 0.60 | 0.70 | 0.77 | 0.89 |

Determine the concentration of silver in the sample, and obtain 95% confidence limits for this concentration.

Equations (4.4) and (4.5) yield $a = 0.3218$ and $b = 0.0186$. The ratio of these figures gives the silver concentration in the test sample as $17.3\,\mu\mathrm{g/ml}$. The confidence limits for this result can be determined with the aid of Eq. (4.12). Here $s_{y/x}$ is 0.01094, $\bar{y} = 0.6014$, and $\sum_i (x_i - \bar{x})^2 = 700$. The value of $s_{x_E}$ is thus 0.749 and the confidence limits are $17.3 \pm 2.57 \times 0.749$, i.e. $17.3 \pm 1.9\,\mu\mathrm{g/ml}$.

Although it is an elegant approach to the common problem of matrix interference effects, the method of standard additions has a number of disadvantages: it is difficult to automate, and it may use larger quantities of sample than other methods. In statistical terms its principal disadvantage is that it is an extrapolation method, and is thus less precise than interpolation techniques. In the example just given, it is easy to show that, if an unknown concentration of silver added to the sample of photographic waste yielded an absorbance value of 0.65, the concentration of the added silver would be $17.6\,\mu\mathrm{g/ml}$, with confidence limits of $17.6 \pm 1.6\,\mu\mathrm{g/ml}$. Comparing this result with that obtained in the example, the improved confidence limits are apparent. As we have seen, the confidence limits for a point on a regression line vary with the $y$-value, and are at a minimum when $y = \bar{y}$ (see above, Section 4.6). In practice, as shown by both this example and the example in Section 4.6, this variation in the confidence limits with change in $y$ may not be very great, so the confidence limits are by no means bad even for a result derived from a substantial extrapolation.

## 4.9 USE OF REGRESSION LINES FOR COMPARING ANALYTICAL METHODS

If an analytical chemist develops a new method for the determination of a particular analyte, he will certainly wish to validate it by (amongst other techniques) applying it to a series of samples already studied by use of another reputable or standard procedure. In making such a comparison, the principal interest will be the identification of systematic errors — does the new method give results that are significantly higher or lower than the established procedure? In cases where an analysis is repeated several times over a very limited concentration range, such a comparison can be made by using the statistical tests described in the previous chapter (see Sections 3.2 and 3.3). Such procedures will not be appropriate in instrumental analyses, which are often used over large concentration ranges.

When two methods are to be compared at different analyte concentrations, the procedure illustrated in Fig. 4.10 is normally adopted. One axis of a regression graph is used for the results obtained by the new method, and the other axis for the results obtained by applying the reference or comparison method to the same samples. (The important question of allocation of an axis to a method is further discussed below.) Each point on the graph thus represents a single sample analysed by two separate methods. The methods of the preceding sections are then applied to calculate the slope ($b$), the intercept ($a$) and the product-moment correlation coefficient ($r$) of the regression line. It is clear that if each sample yields an identical result with both analytical methods the regression line will have a zero intercept, and a slope and a correlation coefficient of 1 (Fig. 4.10a).

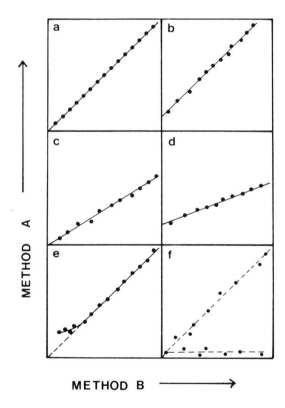

METHOD B ⟶

Fig. 4.10 – Use of a regression line to compare two analytical methods: (a) shows perfect agreement between the two methods for all the samples; (b)-(f) illustrate the results of various types of systematic error (see text).

In practice, of course, this never occurs — even if systematic errors are entirely absent, random errors ensure that the two analytical procedures will not give results in exact agreement for all the samples.

Deviations from the 'ideal' situation ($a = 0$, $b = r = 1$) can occur in a number of different ways. First, it is possible that the regression line will have a slope of 1, but a non-zero intercept. That is, one method of analysis may yield a result higher or lower than the other by a fixed amount. Such an error might occur if the background signal for one of the methods was wrongly calculated (Fig. 4.10b). A second possibility is that the slope of the regression line is $>1$ or $<1$, indicating that a systematic error may be occurring in the slope of one of the individual calibration plots (Fig. 4.10c). These two errors may of course occur simultaneously (Fig. 4.10d). Further possible types of systematic error are revealed if the plot is curved (Fig. 4.10e). Speciation problems may give surprising results (Fig. 4.10f). This type of plot might arise if an analyte occurred in two chemically distinct forms, the proportions of which varied from sample to sample. One of the methods under study (here plotted on the $y$-axis) might detect only one form of the analyte, while the second method detected both forms.

In practice, the analyst most commonly wishes to test for an intercept differing significantly from zero, and a slope differing significantly from 1. Such tests are performed by determining the confidence limits for $a$ and $b$, generally at the 95% significance level. The calculation is thus in practice very similar to that described in Section 4.5, and is most usefully performed by using a simple computer program.

*Example*. The level of lead in ten fruit juice samples was determined by a new potentiometric stripping analysis (PSA) method employing a glassy-carbon working electrode, and the results were compared with those obtained by using a flameless atomic-absorption spectrometry (AAS) technique. The following data were obtained (all results in $\mu$g/l.).

| Sample | 1 | 2 | 3 | 4 | 5 | 6 | 7 | 8 | 9 | 10 |
|---|---|---|---|---|---|---|---|---|---|---|
| AAS result | 35 | 75 | 75 | 80 | 125 | 205 | 205 | 215 | 240 | 350 |
| PSA result | 35 | 70 | 80 | 80 | 120 | 200 | 220 | 200 | 250 | 330 |

(S. Mannino, *Analyst*, 1982, **107**, 1466).

These results are plotted on a regression line (Fig. 4.11), with the AAS results assigned to the $x$-axis, and the PSA results to the $y$-axis. Using the methods of the preceding sections it may readily be shown that:

$a = 3.87$; $b = 0.963$; $r = 0.9945$

Further calculations show that:

$s_{y/x} = 10.56$; $s_a = 6.64$; $s_b = 0.0357$

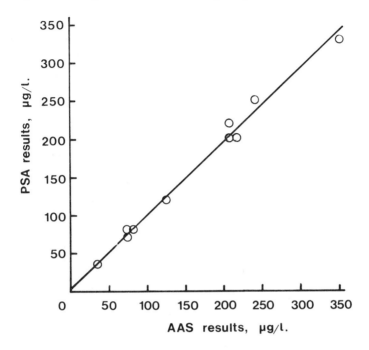

Fig. 4.11 – Comparison of two analytical methods: the plot shows the data on p. 104.

and the use of the appropriate $t$-value for 8 degrees of freedom ($t = 2.31$) gives the 95% confidence limits for the intercept and slope as:

$$a = 3.87 \pm 15.34 \text{ and } b = 0.963 \pm 0.083$$

From these figures it is clear that the calculated slope and intercept do not differ significantly from the "ideal" values of 1 and 0 respectively, and thus that there is no evidence for systematic differences between the two sets of results.

Two further points may be mentioned in connection with this example. First, inspection of the literature of analytical chemistry shows that authors frequently place great stress on the value of the correlation coefficient in such comparative studies. In the example, however, it played no direct role in establishing whether or not systematic errors had occurred. Even if the regression line had been slightly curved, the correlation coefficient might still have been close to 1 (cf. Section 4.3 above). It is thus evident that the calculation of $r$ is of less importance in the present context than the establishment of confidence limits for the slope and the intercept. In some cases it may be found that the $r$ value is not very close to 1, even though the slope and the intercept are not significantly different from 1 and 0 respectively. Such a result would suggest very poor precision for either one or both of the methods under study. The precision of the two methods can, of course, be determined and compared by the methods of Chapters 2 and 3. In practice it is desirable that this should be done *before* the regression line comparing the methods is plotted — the reason for this is explained below. The second point to note is that, although it is clearly desirable to compare the methods over as wide a range of concentrations as possible, it may not in practice be feasible to obtain real samples which have analyte concentrations that are evenly spaced in the range. In the example given, the fruit juices examined tended to fall into groups with fairly similar lead concentrations, so samples with lead levels between ca. 130 and 200 $\mu g/l$. and between 260 and 320 $\mu g/l$. could not be studied. This point will also be referred to below.

Although almost universally adopted in comparative studies of instrumental methods, the approach described here is open to serious theoretical objections on several grounds. First, as has been emphasized throughout this chapter, the line of regression of $y$ on $x$ is calculated on the assumption that the errors in the $x$-values are negligible — all the errors are assumed to occur in the $y$-direction. While generally valid in the production of a calibration plot for a single analytical method, this assumption can evidently not be justified when the regression line is used for comparison purposes. In such comparisons it can be taken as certain that random errors will occur in both analytical methods, i.e. in both the $x$ and $y$ directions. This would suggest that the equations used to calculate the regression line itself may not be valid. Practical tests and simulations, however, show that the present simple approach does give surprisingly reliable results, provided that three conditions are fulfilled.

1. The more precise method is plotted on the $x$-axis: this is the reason for making preliminary investigations on the precisions of the two methods — see above.

2. A reasonable number (10 at least, as in the example above) of points are plotted in the comparison. Since the confidence limit calculations are based on $(n - 2)$ degrees of freedom, it is particularly important to avoid small values of $n$.

3. The experimental points should cover the concentration range of interest in a roughly uniform fashion: as we have seen, this requirement may be difficult to fulfil in comparative studies on real samples.

There is a second theoretical objection to using the line of regression of $y$ on $x$, as calculated in Sections 4.4 and 4.5, in the comparison of two analytical methods. This regression line assumes not only that the $x$-direction errors are zero: it also assumes that the error in the $y$-values is *constant,* i.e. it does not vary with concentration, and that all the points thus have equal weighting when the slope and intercept of the line are calculated. This assumption is obviously likely to be invalid in practice. In many analyses, the relative standard deviation (coefficient of variation) is roughly constant over a range of concentrations: the absolute error thus increases with the concentration of the analyte, rather than having the same value at all concentrations. It follows that 'unweighted' regression lines are also of very questionable validity in other situations, e.g. when applied to calibration plots for a single analytical procedure. In principle, weighted regression lines should be used instead. This concept is elaborated in the next section. Meanwhile, it can be re-affirmed that, despite the theoretical objections, unweighted regression lines provide useful information in comparative studies provided that the requirements listed above are fulfilled.

## 4.10 WEIGHTED REGRESSION LINES

The comments made in the previous section on conventional or unweighted regression calculations indicate that weighted regression calculations should perhaps be adopted far more frequently than is in fact the case. The calculations involved in using weighted regression methods are only a little more complicated than those of the previous sections, and can easily be performed on a micro-computer, but they require additional information on the errors occurring at different concentration levels, or at least the formulation of additional assumptions about such errors. These considerations perhaps explain why unweighted regression calculations are less used than they might be. In this section the application of weighted regression methods is outlined. It is assumed that the weighted regression line is to be used for the determination of a single analyte rather than for the comparison of two separate methods.

Let us consider in more detail the simple situation that arises when the error in a regression calculation is approximately proportional to the concentration of the analyte. When the errors at different points on the calibration graph are expressed by 'error bars' (Fig. 4.12) the bars thus get larger as the concentration increases. In this situation it is evident that the regression line must be calculated to give additional weight to those points where the error bars are smallest: it is more important for the calculated line to pass close to such points than to pass close to the points representing higher concentrations with the largest errors. This result is achieved by giving each point a weighting inversely proportional to the corresponding variance, $s^2$; this logical procedure is of general application. Thus, if the individual points are denoted by $(x_1, y_1)$, $(x_2, y_2)$ etc. as usual, and the corresponding standard deviations are $s_1, s_2$ etc., then the individual weights, $w_1, w_2$ etc. are given by:

$$w_i = s_i^{-2} / (\sum_i s_i^{-2} / n) \tag{4.13}$$

It will be seen that the weights have been scaled so that their sum is equal to the number of points on the graph: this simplifies the subsequent calculations.

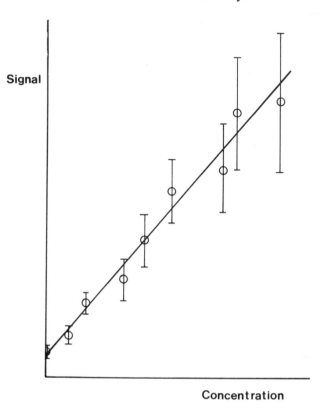

Fig. 4.12 – The weighting of errors in a regression calculation.

The slope and the intercept of the regression line are then given by:

$$b = \frac{\sum_i w_i x_i y_i - n\bar{x}_w\bar{y}_w}{\sum_i w_i x_i^2 - n\bar{x}_w^2} \tag{4.14}$$

and

$$a = \bar{y}_w - b\bar{x}_w \tag{4.15}$$

In Eq. (4.15), $\bar{x}_w$ and $\bar{y}_w$ represent the coordinates of the **weighted centroid**, $(\bar{x}_w, \bar{y}_w)$ through which the weighted regression line must pass. These coordinates are given, as expected, by $\bar{x}_w = \sum_i w_i x_i/n$ and $\bar{y}_w = \sum_i w_i y_i/n$.

*Example.* Calculate the unweighted and weighted regression lines for the following calibration data. For each line calculate also the concentrations of test samples with absorbances of 0.100 and 0.600.

| Concentration, $\mu$g/ml | 0 | 2 | 4 | 6 | 8 | 10 |
|---|---|---|---|---|---|---|
| Absorbance | 0.009 | 0.158 | 0.301 | 0.472 | 0.577 | 0.739 |
| Standard deviation | 0.001 | 0.004 | 0.010 | 0.013 | 0.017 | 0.022 |

Application of Eqs. (4.4) and (4.5) shows that the slope and intercept of the *unweighted* regression line are respectively 0.0725 and 0.0133. The concentrations corresponding to absorbances of 0.100 and 0.600 are thus readily found to be 1.20 and 8.09 $\mu$g/ml respectively.

The *weighted* regression line is a little harder to calculate: in the absence of a suitable computer program it is usual to set up a table as follows.

| $x_i$ | $y_i$ | $s_i$ | $1/s_i^2$ | $w_i$ | $w_i x_i$ | $w_i y_i$ | $w_i x_i y_i$ | $w x_i^2$ |
|---|---|---|---|---|---|---|---|---|
| 0 | 0.009 | 0.001 | 1000000 | 5.535 | 0 | 0.0498 | 0 | 0 |
| 2 | 0.158 | 0.004 | 62500 | 0.346 | 0.692 | 0.0547 | 0.1093 | 1.384 |
| 4 | 0.301 | 0.010 | 10000 | 0.055 | 0.220 | 0.0166 | 0.0662 | 0.880 |
| 6 | 0.472 | 0.013 | 5917 | 0.033 | 0.198 | 0.0156 | 0.0935 | 1.188 |
| 8 | 0.577 | 0.017 | 3460 | 0.019 | 0.152 | 0.0110 | 0.0877 | 1.216 |
| 10 | 0.739 | 0.022 | 2066 | 0.011 | 0.110 | 0.0081 | 0.0813 | 1.100 |
| | | SUMS: | 1083943 | 5.999 | 1.372 | 0.1558 | 0.4380 | 5.768 |

From these figures it is clear that $\bar{y}_w = 0.1558/6 = 0.0260$ and $\bar{x}_w = 1.372/6 = 0.229$. By Eq. (4.14), $b$ is calculated from

$$b = \frac{0.438 - (6 \times 0.229 \times 0.026)}{5.768 - [6 \times (0.229)^2]} = 0.0738$$

so $a$ is given by $0.0260 - (0.0738 \times 0.229) = 0.0091$.

These values for $a$ and $b$ can be used to show that absorbance values of 0.100 and 0.600 correspond to concentrations of 1.23 and 8.01 $\mu$g/ml respectively.

Careful comparison of the results of the unweighted and weighted regression calculations is very instructive. The effects of the weighting process are clear. The weighted centroid $(\bar{x}_w, \bar{y}_w)$ is much closer to the origin of the graph than the unweighted centroid $(\bar{x}, \bar{y})$, and the weighting given to the points nearer the origin — and particularly to the first point (0, 0.009), which has the smallest error — ensures that the weighted regression line has an intercept very close to this point. The slope and intercept of the weighted line are remarkably similar to those of the unweighted line, however, with the result that the two methods give very similar values for the concentrations of samples having absorbances of 0.100 and 0.600. It must not be supposed that these similar values arise simply because in this example the experimental points fit a straight line very well. In practice the weighted and unweighted regression lines derived from a set of experimental data have similar slopes and intercepts even if the scatter of the points about the line is substantial.

We might thus be forgiven for thinking that weighted regression calculations have little to recommend them. They require more information (in the form of estimates of the standard deviation at various points on the graph), and are far more complex to execute, but they seem to provide data that are remarkably similar to those obtained from the much simpler unweighted regression method. Such considerations may indeed account for the widespread neglect of weighted regression calculations in practice. But an analytical chemist using instrumental methods does not employ regression calculations simply to determine the slope and intercept of the calibration plot and the concentrations of test samples. He also wants to obtain estimates of the errors or confidence limits of those concentrations and it is in this context that the weighted regression method provides much more realistic results. In Section 4.6 we used Eq. (4.9) to estimate the standard deviation $(s_{x_0})$ and hence the confidence intervals of a concentration calculated by using a single $y$-value and an unweighted regression line. Application of this equation to the data in the example above shows that the unweighted confidence limits for the solutions having absorbances of 0.100 and 0.600 are $1.20 \pm 0.65$ and $8.09 \pm 0.63$ $\mu$g/ml respectively. As in the example in Section 4.6, these confidence intervals are very similar. In the present example, however, such a result is entirely unrealistic. The experimental data show that the errors of the observed $y$-values increase as $y$ itself increases, the situation expected for a method having a roughly constant relative standard deviation. We would expect that this increase in $s_i$ with increasing $y$ would also be reflected in the confidence limits of the determined concentrations: the confidence limits for the solution with an absorbance of 0.600 should be much greater (i.e. worse) than those for the solution with an absorbance of 0.100.

In weighted regression calculations, the standard deviation $(s_{x_0 w})$ of a predicted concentration is given by:

$$s_{x_0 w} = \frac{s_{(y/x)w}}{b} \left\{ \frac{1}{w_0} + \frac{1}{n} + \frac{(y_0 - \bar{y}_w)^2}{b^2 (\sum_i w_i x_i^2 - n\bar{x}_w^2)} \right\}^{\frac{1}{2}} \qquad (4.16)$$

In this equation, $s_{(y/x)w}$ is given by:

$$s_{(y/x)w} = \left\{ \frac{(\sum_i w_i y_i^2 - n\bar{y}_w^2) - b^2 (\sum_i w_i x_i^2 - n\bar{x}_w^2)}{n - 2} \right\}^{\frac{1}{2}} \qquad (4.17)$$

and $w_0$ is a weighting appropriate to the value of $y_0$. Equation (4.16) is clearly similar in form to Eq. (4.9). It confirms that points close to the origin, where the weights are highest, and points near the centroid, where $y_0 - \bar{y}_w$ is small, will have the smallest confidence limits (Fig. 4.13). The major difference between the two equations is the term $1/w_0$ in Eq. (4.16). Since $w_0$ falls sharply as $y$

**Signal**

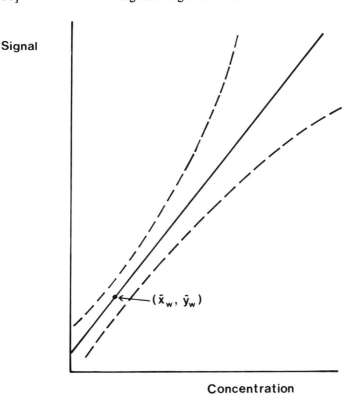

$(\bar{x}_w, \bar{y}_w)$

**Concentration**

Fig. 4.13 – General form of the confidence limits for a concentration determined by using a weighted regression line.

increases, this term ensures that the confidence limits increase with increasing $y_0$, as we expect.

Application of Eq. (4.16) to the data in the example above shows that the test samples with absorbance of 0.100 and 0.600 have confidence intervals for the calculated concentrations (1.23 and 8.01 $\mu$g/ml) of ±0.12 and ±0.72 $\mu$g/ml respectively. We note that these confidence intervals are proportional to the observed absorbances of the two solutions. In addition the confidence interval for the less concentrated of the two samples is smaller than in the unweighted regression calculation, while for the more concentrated sample the opposite is true. All these results accord much more closely with the reality of a calibration experiment than do the results of the unweighted regression calculation.

We therefore conclude that, although weighted regression calculations are more complicated to perform than their unweighted counterparts, they give much more realistic results for the confidence limits of predicted concentrations in conventional instrumental analysis. With the advent of suitable computer programs to simplify the calculations, the use of weighted methods should be encouraged.

## 4.11 CURVILINEAR REGRESSION

Thus far, our discussion of calibration methods has been confined to experiments where we can assume that the instrument response is proportional to the analyte concentration. This assumption is usually valid, because analytical chemists have always favoured such methods, and taken experimental precautions to ensure that linearity of response is preserved over as wide a concentration range as possible. Examples of such precautions include the control of the emission line-width of a hollow-cathode lamp in atomic-absorption spectrometry, and the positioning of the sample cuvette to minimize inner filter effects in molecular fluorescence spectrometry. Despite such endeavours, however, many analytical methods give curved calibration plots over part of the concentration range of interest. Particularly common is the situation where the calibration plot is linear (or approximately so) at low analyte concentrations, but becomes curved at higher concentrations. In other cases, for example in immunoassays, and in the response of some ion-selective electrodes, the calibration plot is obviously curved at all concentrations. When calibration plots are curved we shall still need answers to the questions listed in Section 4.2, but the questions will pose much more formidable statistical problems than do linear calibration experiments. The complete solution of these problems is certainly beyond the scope of this book, but it is worth giving them some preliminary attention. The reader is referred to the bibliography for more complete treatments.

The first question to be examined is — how do we detect curvature in a calibration plot? That is, how do we distinguish between a plot which is best fitted by a straight line, and one which is best fitted by a gentle curve? Since the degree of curvature may be small, and/or occur over only part of the plot, this is clearly not a trivial question. Moreover, despite its widespread use for testing the goodness of fit of linear graphs, the product–moment correlation coefficient ($r$) is of little value in testing for curvature: we have seen (Section 4.3) that lines with obvious curvature may still give very high $r$ values. An analyst would naturally hope that any test for curvature could be applied fairly easily in routine work without extensive calculations. Several such tests are available, based on the use of $y$-residuals on the calibration plot.

As we have seen (Section 4.5) a $y$-residual, $y_i - \hat{y}$, represents the difference between an experimental value of $y$ and the $\hat{y}$ value calculated from the regression line at the same value of $x$. If a linear calibration plot is appropriate, and if the random errors in the $y$-values are normally distributed, the residuals themselves should be normally distributed about the value zero. If this turns out not to be true in practice, then (neglecting the possibility that the experimental errors themselves are not normally distributed) we must suspect that the fitted regression

line is not of the correct type. In the worked example given in Section 4.5, the
$y$-residuals were shown to be $+0.58$, $-0.38$, $-0.24$, $-0.50$, $+0.34$, $+0.18$, and
$+0.02$. It is clear that these values sum to zero (allowing for possible rounding
errors, this must always be true), and are approximately symmetrically distributed about 0. Although it is impossible to be certain, especially with small
numbers of data points, that these residuals are normally distributed, there is
certainly no contrary evidence in this case, i.e. no evidence to support a non-
linear calibration plot.

A second test suggests itself on inspection of the signs of the residuals
given above. As we move along the calibration plot, i.e. as $x$ increases, positive
and negative residuals will be expected to occur in random order if the data
are well fitted by a straight line. If, in contrast, we attempt to fit a straight
line to a series of points that actually lie on a smooth curve, then the signs of
the residuals will no longer have a random order, but will occur in *sequences* of
positive and negative values. Examining again the residuals given above, we find
that the order of signs is $+ - - - - + +$. To test whether these sequences
of $+$ and $-$ residuals indicate the need for a non-linear regression line, we need
to know the probability that such an order could occur by chance; such calculations are described in the next chapter. Unfortunately the small number of data
points makes it quite likely that these and other sequences could indeed occur
by chance, so any conclusions drawn must be treated with great caution. The
choice between straight line and curvilinear regression methods is probably best
made by using the curve-fitting techniques outlined below.

In the situation where a calibration plot is linear over part of its range
and curved elsewhere, it is of great importance for the analytical chemist to
establish the range over which linearity can be assumed. Possible approaches
to this problem are outlined in the following example.

*Example.* Investigate the linear calibration range of the following fluorescence
experiment.

| Fluorescence intensity | 0.1 | 8.0 | 15.7 | 24.2 | 31.5 | 33.0 |
|---|---|---|---|---|---|---|
| Concentration, $\mu$g/ml | 0 | 2 | 4 | 6 | 8 | 10 |

Inspection of the data shows that the part of the graph near the origin
corresponds rather closely to a straight line with a near-zero intercept and a
slope of about 4. The fluorescence of the 10-$\mu$g/ml standard solution is
clearly lower than would be expected on this basis, and there is some
possibility that the departure from linearity has also affected the fluorescence
of the 8-$\mu$g/ml standard. We first apply (unweighted) linear regression
calculations to all the data. Application of the methods of Sections 4.3 and
4.4 gives the results $a = 1.357$, $b = 3.479$ and $r = 0.9878$. Again we recall
that the high value for $r$ may be deceptive, though it may be used in a
comparative sense (see below). The $y$-residuals are found to be $-1.257$,
$-0.314$, $+0.429$, $+1.971$, $+2.314$, and $-3.143$, with the sum of squares of

the residuals equal to 20.981. The trend in the values of the residuals suggests that the last value in the table is probably outside the linear range.

We confirm this suspicion by applying the linear regression equations to the first five points only. This gives $a = 0.100$, $b = 3.950$ and $r = 0.9998$. The slope and intercept are much closer to the values expected for the part of the graph closest to the origin, and the $r$ value is higher than in the first calculation. The residuals of the first five points from this second regression equation are 0, 0, $-0.2$, $+0.4$ and $-0.2$, with a sum of squares of only 0.24. Use of the second regression equation shows that the fluorescence expected from a 10-$\mu$g/ml standard is 39.6, i.e. the residual is $-6.6$. Use of a $t$-test (Chapter 3) would show that this last residual is significantly greater than the average of the other residuals: alternatively a test could be applied (Section 3.6) to demonstrate that it is an 'outlier' amongst the residuals. In this example, such calculations are hardly necessary: the enormous residual for the last point, coupled with the very low residuals for the other five points and the greatly reduced sum of squares, confirms that the linear range of the method does not extend as far as 10 $\mu$g/ml.

Having established that the last data point can be excluded from the linear range, we can repeat the process to study the point (8, 31.5). We do this by calculating the regression line for only the first four points in the table, with the results $a = 0$, $b = 4.00$, $r = 0.9998$. The correlation coefficient value suggests that this line is about as good a fit of the points as the previous one, in which five points were used. The residuals for this third calculation are $+0.1$, 0, $-0.3$, and $+0.2$, with a sum of squares of 0.14. With this calibration line the $y$-residual for the 8-$\mu$g/ml solution is $-0.5$: this value is larger than the other residuals but probably not by a significant amount. It can thus be concluded that it is reasonably safe to include the point (8, 31.5) within the linear range of the method. In making a marginal decision of this kind, the analytical chemist will take into account the accuracy required in his results, and the reduced value of a method for which the calibration range is very short. The calculations described above are summarized in Fig. 4.14.

Once a decision has been taken that a set of calibration points cannot be satisfactorily fitted by a straight line, the analyst can play one further card before resigning himself to the complexities of curvilinear regression calculations. He may be able to **transform** the data so that a non-linear relationship is changed into a linear one. A particularly common method is to plot log $y$ and/or log $x$ instead of $y$ and $x$. This will produce linear relationships from curves orginally of a form such as $y = px^q$. Such transformations are quite regularly applied to the results of certain analytical methods — immunoassays, for example. It is important to note that the transformations may also affect the nature of the errors at different points on the calibration plot. Data with errors that increase as $x$

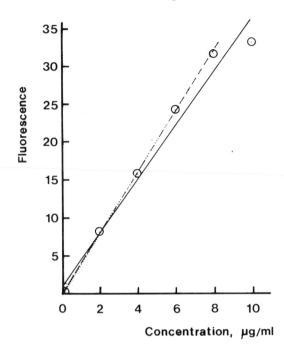

Fig. 4.14 – Curvilinear regression: identification of the linear range. The data on p. 113 are used; the unweighted linear regression lines through all the points (——), through the first five points only (– – – – –), and through the first four points only (. . . . .) are shown.

increases may be transformed into data with more or less constant errors. Sometimes transformations are performed with the specific intention of affecting the error distribution rather than the shape of the curve itself: this is because some of the simpler curve-fitting methods assume that the errors do not vary with $x$.

The fitting of a curve to a set of calibration points is a complex iterative process, best achieved with the aid of an interactive computer program that displays the results of the fitting attempts on a visual display unit or printer for evaluation by the analyst. The program will seek to fit a curve of the general form $y = a + bx + cx^2 + dx^3 + \dots$, i.e. a polynomial in $x$. The success of the fitting process will be assessed, exactly as in the linear regression calculations, by the least-squares method; the program will look for the curve that minimizes the sum of the squares of the $y$-residuals. Programs that allow for the different points to have different weights are available, but are inevitably more complex than those that assume a constant error over the whole calibration curve. Such computer methods must be used with common sense. If there are $n$ points on the calibration graph, the highest order polynomial that need be examined is that with order $(n - 1)$; but in general the lowest-order polynomial giving a good fit will be sought, and quadratic and cubic equations often provide satisfactory results.

If the curvature of the regression plot is not too severe, and provided that the calibration points are not too widely separated — conditions usually satisfied in practical analytical work — a simple but approximate method can be used in place of the complex curve-fitting approach. This involves drawing a straight line between each pair of points and calculating concentrations by using linear interpolation, i.e. the curve is treated as a series of short straight segments. This method does not, of course, furnish confidence limits for the calculated concentrations, but its simplicity is very attractive and the systematic errors involved are often surprisingly small. This point is illustrated in one of the exercises at the end of the chapter.

## BIBLIOGRAPHY

F. S. Acton, *Analysis of Straight Line Data,* Dover, New York, 1966. A well-written and comprehensive treatment, with plenty of examples, including chemical ones. Despite its title, it also gives some attention to curvilinear regression.

R. Caulcutt and R. Boddy, *Statistics for Analytical Chemists,* Chapman & Hall, London, 1983. An entirely pragmatic approach, lacking statistical theory but full of commonsense examples. Oriented towards industrial analytical chemistry.

O. L. Davies and P. L. Goldsmith, *Statistical Methods in Research and Production,* Longmans, London, 1982. A fairly full treatment of regression and correlation methods, with considerable emphasis on advanced concepts such as curvilinear regression and multiple regression problems.

N. R. Draper and H. Smith, *Applied Regression Analysis,* Wiley, New York, 1966. An established work in this field, with a comprehensive coverage.

H. Kaiser and A. C. Menzies, *The Limit of Detection of a Complete Analytical Procedure,* Hilger, London, 1968. An interesting discussion of detection limits.

G. W. Snedecor and W. G. Cochran, *Statistical Methods,* Iowa State University, 1967. Gives an excellent general account of regression and correlation procedures.

## EXERCISES

1. In a laboratory containing polarographic equipment six samples of dust were taken at various distances from the polarograph and the mercury content of each sample was determined. The following results were obtained.

| Distance from polarograph, m: | 1.4 | 3.8 | 7.5 | 10.2 | 11.7 | 15.0 |
|---|---|---|---|---|---|---|
| Mercury concentration, ng/g: | 2.4 | 2.5 | 1.3 | 1.3 | 0.7 | 1.2 |

Examine the possibility that the mercury contamination arose from the polarograph.

2. The response of a colorimetric test for glucose was checked with the aid of standard glucose solutions. Determine the correlation coefficient from the following data, and comment on the result.

| Glucose concentration, m$M$: | 0 | 2 | 4 | 6 | 8 | 10 |
|---|---|---|---|---|---|---|
| Absorbance: | 0.002 | 0.150 | 0.294 | 0.434 | 0.570 | 0.704 |

3. The following results were obtained when each of a series of standard silver solutions was analysed by flame atomic-absorption spectrometry.

| Concentration, ng/ml: | 0 | 5 | 10 | 15 | 20 | 25 | 30 |
|---|---|---|---|---|---|---|---|
| Absorbance: | 0.003 | 0.127 | 0.251 | 0.390 | 0.498 | 0.625 | 0.763 |

Determine the slope and intercept of the calibration plot, along with their confidence limits.

4. Using the data of exercise 3, estimate the confidence limits for the silver concentrations in (a) a sample giving an absorbance of 0.456 in a single determination, and (b) a sample giving absorbance values of 0.308, 0.314, 0.347, and 0.312 in four separate analyses.

5. Estimate the limit of detection of the silver analysis from the data in exercise 3.

6. The gold content of a concentrated sea-water sample was determined by using atomic-absorption spectrometry with the method of standard additions. The results obtained were as follows.

| Gold added, ng per ml of concentrated sample: | 0 | 10 | 20 | 30 | 40 | 50 | 60 | 70 |
|---|---|---|---|---|---|---|---|---|
| Absorbance: | 0.257 | 0.314 | 0.364 | 0.413 | 0.468 | 0.528 | 0.574 | 0.635 |

Estimate the concentration of the gold in the concentrated sea-water, and determine confidence limits for this concentration.

7. The fluorescence of each of a series of acidic solutions of quinine was determined 5 times; the results are given below.

| Concentration, ng/ml: | 0 | 10 | 20 | 30 | 40 | 50 |
|---|---|---|---|---|---|---|
| Fluorescence intensity | 4 | 22 | 44 | 60 | 75 | 104 |
| (arbitrary units). | 3 | 20 | 46 | 63 | 81 | 109 |
| | 4 | 21 | 45 | 60 | 79 | 107 |
| | 5 | 22 | 44 | 63 | 78 | 101 |
| | 4 | 21 | 44 | 63 | 77 | 105 |

Determine the slope and intercept of (a) the unweighted regression line, and (b) the weighted regression line. Calculate, using both regression lines, the confidence limits for the concentrations of solutions with fluorescence intensities of 15 and 90 units.

8. An ion-selective electrode (ISE) determination of sulphide from sulphate-reducing bacteria was compared with a gravimetric determination. The results obtained were expressed in milligrams of sulphide.

| Sample: | 1 | 2 | 3 | 4 | 5 | 6 | 7 | 8 | 9 | 10 |
|---|---|---|---|---|---|---|---|---|---|---|
| Sulphide (ISE method): | 108 | 12 | 152 | 3 | 106 | 11 | 128 | 12 | 160 | 128 |
| Sulphide (gravimetry): | 105 | 16 | 113 | 0 | 108 | 11 | 141 | 11 | 182 | 118 |

Comment on the suitability of the ISE method for this sulphide determination. (I. K. Al-Hitti, G. J. Moody and J. D. R. Thomas, *Analyst,* 1983, **108,** 43).

9. In the determination of lead in aqueous solution by electrochemical atomic-absorption spectrometry with graphite-probe atomization, the following results were obtained.

| Lead concentration, ng/ml: | 10 | 25 | 50 | 100 | 200 | 300 |
|---|---|---|---|---|---|---|
| Absorbance: | 0.05 | 0.17 | 0.32 | 0.60 | 1.07 | 1.40 |

Investigate the linear calibration range of this experiment. (Based on S. K. Giri, C. K. Shields, D. Littlejohn and J. M. Ottaway, *Analyst,* 1983, **108,** 244).

10. In an instrumental analysis the following calibration data were obtained (arbitrary units).

| Concentration: | 0 | 1 | 2 | 3 | 4 | 5 | 6 | 7 | 8 | 9 | 10 |
|---|---|---|---|---|---|---|---|---|---|---|---|
| Signal: | 0.2 | 3.6 | 7.5 | 11.5 | 15.0 | 17.0 | 20.4 | 22.7 | 25.9 | 27.6 | 30.2 |

Draw the calibration plot, and compare the concentrations corresponding to signals of 5, 16 and 27 units, determined by (a) fitting a straight line to the points, (b) fitting the curve $y = 4x - 0.1x^2$, (c) treating the curve as a series of linear segments between $(x_1, y_1)$ and $(x_2, y_2)$, $(x_2, y_2)$ and $(x_3, y_3)$ etc., and (d) treating the curve as a series of linear segments but using only the points $(x_1, y_1)$, $(x_3, y_3)$, $(x_5, y_5)$ etc.

# 5

# Rapid and non-parametric methods

## 5.1 INTRODUCTION

This chapter describes a series of useful statistical methods that differ in one important respect from the methods so far discussed. The tests developed in the previous chapters have all assumed that the data being examined follow the normal (Gaussian) distribution. The validity of this assumption is supported by the central limit theorem (Chapter 2), which shows that the sampling distribution of the mean may be approximately normal even if the parent population has quite a different distribution, the closeness of the approximation improves as the size of the sample increases.

There are several reasons, however, why we should be interested in methods that do not require such an assumption. First, some sets of data that are of interest to analytical chemists are not normally distributed. For example (cf. Chapter 2) the concentrations of antibody in the blood sera of a group of different people can be expressed approximately as a log-normal distribution. Secondly, in many experiments we use very small samples of data indeed (e.g. only 3 or 4 titrations in a volumetric analysis), so that even the central limit theorem will be of little reassurance. Finally, the methods of the preceding chapters often require quite a lot of detailed calculation; by contrast many of the methods to be described in this chapter greatly simplify the calculation steps. Indeed some of the tests are easily performed mentally, a very convenient feature.

Statistical methods that make no assumptions about the shape of the distribution from which the data are taken are called **non-parametric** or distribution-free methods. In subsequent sections we shall discuss a number of non-parametric methods, being content in most cases to describe the operation of the tests without detailed theoretical discussion.

## 5.2 THE MEDIAN

In previous chapters we have used the arithmetic mean or average as the 'measure of central tendency' of a set of results. In non-parametric statistics, however, the **median** is usually used instead. In order to calculate the median of $n$ observations, we arrange them in ascending order (in the unlikely event that the number of observations is very large, this sorting process can be performed very quickly by simple programs available for most microcomputers). The median is then the value of the $(n + 1)$th observation if $n$ is odd: and the average of the $\frac{1}{2}n$th and the $(\frac{1}{2}n + 1)$th observations if $n$ is even. Determining the median of a set of experimental results thus normally requires little or no calculation. Moreover, in many cases it may be a more realistic measure of central tendency than the arithmetic mean.

*Example.* Determine the mean and the median for the following four titration values.

$$25.01, \quad 25.04, \quad 25.06, \quad 25.21 \text{ ml}$$

It is easy to calculate that the mean of these four observations is 25.08 ml, and that the median — in this case the average of the 2nd and 3rd values, the observations already being in numerical order — is 25.05 ml. The mean is greater than any of the three closely-grouped values (25.01, 25.04 and 25.06 ml) and may thus be a less realistic measure of central tendency than the median. Instead of calculating the median we could use the methods of Chapter 3 to test the value 25.21 as a possible outlier, and calculate the mean according to the result obtained, but this approach will *assume* that the data come from a normal population.

This simple example illustrates one valuable property of the median: it is not affected by outlying values. Confidence limits (cf. Chapter 2) for the median can be estimated with the aid of the binomial distribution. This calculation can be performed even when the number of measurements is small, but is not likely to be required in analytical chemistry, where the median is generally used only as a rapid estimate of an average. The reader is referred to the bibliography for further information.

In non-parametric statistics the usual measure of dispersion (replacing the standard deviation) is the **interquartile range**. As we have seen, the median divides the sample of measurements into two equal halves: if each of these halves is further divided into two the points of division are called the **lower** and **upper quartiles**. Several different conventions are used in making this calculation, and the interested reader should again consult the bibliography. The interquartile range is not widely used in analytical work, but various statistical tests can be performed on it.

## 5.3 THE SIGN TEST

The **sign test** is amongst the simplest of all non-parametric methods, and was first discussed in the early eighteenth century. It can be used in a number of ways, the simplest of which is demonstrated by the following example.

> *Example.* A pharmaceutical preparation is claimed to have a median content of 8% of a particular constituent. Successive batches were found in practice to contain 7.3, 7.1, 7.9, 9.1, 8.0, 7.1, 6.8 and 7.3% of the constituent. At the 5% significance level, does this indicate that the claimed median is incorrect?

In Chapter 3 (Section 3.2) it was shown that such problems could be tackled by using the $t$-test after calculation of the mean and standard deviation of the experimental data. The $t$-test assumes, however, that the data are normally distributed. The sign test avoids such an assumption, and is much easier to perform. The postulated median is subtracted from each experimental value in turn, and the **sign** of each result is considered. Values equal to the postulated median are ignored entirely. In this case, therefore, we effectively have seven experimental values, six of them lower than the median and hence giving minus signs, and one higher than the median, and thus giving a $+$ sign. To test whether this preponderance of minus signs is significant we use the binomial theorem. This theorem shows that the probability of $r$ out of $n$ signs being minus is given by

$$P(r) = {}^nC_r \cdot p^r q^{(n-r)} \tag{5.1}$$

where ${}^nC_r$ is the number of combinations of $r$ items from a total of $n$ items, $p$ is the probability of getting a minus sign in a single result, and $q$ is the probability of not getting a minus sign in a single result i.e. $q = 1 - p$. Since the median is defined so that half the experimental results lie above it, and half below it, it is clear that if the median *is* 8.0 in this case, then both $p$ and $q$ should be $\frac{1}{2}$. Using Eq. (5.1), we find that $P(6) = {}^7C_6 \times (\frac{1}{2})^6 \times \frac{1}{2} = 7/128$. Similarly we can calculate that the chance of getting 7 negative signs, $P(7)$, is 1/128. Overall, therefore, the probability of getting *6 or more* negative signs in our experiment is 8/128. We are only asking, however, whether the data differ significantly from the postulated median. We must therefore perform a two-tailed test (cf. Chapter 3), i.e. we must calculate the probability of obtaining six or more identical signs (i.e. $\geqslant 6$ plus or $\geqslant 6$ minus signs) when 7 results are taken at random. This is clearly $16/128 = 0.125$. We then compare this figure with 0.05, i.e. we perform our test at the 5% significance level. Since the experimental value is $>0.05$, our null hypothesis, i.e. that the data come from a population with median 8.0, cannot be rejected. As in Chapter 3, it is important to note that we have not proved that the data *do* come from such a population; we have only concluded that such a hypothesis cannot be rejected.

It is apparent from this example that the sign test will involve the frequent use of the binomial distribution with $p = q = \frac{1}{2}$. So common is this approach to non-parametric statistics that most sets of statistical tables include the necessary data, allowing such calculations to be made instantly (see Table A.7). Moreover, in many practical situations, an analyst will always use the same value of $n$, i.e. he will always take the same number of readings or samples, and he will be able to memorize easily the probabilities corresponding to the various numbers of $+$ or $-$ signs.

The sign test can also be used as a non-parametric alternative to the paired $t$-test (Section 3.4) to compare two sets of results for the same samples. Thus, if ten samples are examined by each of two methods, A and B, we can test whether the two methods give significantly different results, by calculating for each sample the difference in the results, i.e. (result obtained by method A – result obtained by method B). The null hypothesis will be that the two methods do not give significantly different results – in practice this will again mean that the probability of obtaining a plus sign (or a minus sign) is 0.5. The number of plus or minus signs actually obtained can be compared with the probability derived from Eq. (5.1). An example of this application of the sign test is given in the exercises at the end of this chapter.

A further use of the sign test is to indicate a trend. This application is illustrated by the following example.

*Example.* The level of a hormone in a patient's blood plasma is measured at the same time each day for 10 days. The resulting data are:

| Day | 1 | 2 | 3 | 4 | 5 | 6 | 7 | 8 | 9 | 10 |
|---|---|---|---|---|---|---|---|---|---|---|
| Level, ng/ml | 5.8 | 7.3 | 4.9 | 6.1 | 5.5 | 5.5 | 6.0 | 4.9 | 6.0 | 5.0 |

Is there any evidence for a trend in the hormone concentration?

Using parametric methods, it would be possible to make a linear regression plot of such data and test whether its slope differed significantly from zero (Chapter 4). Such an approach would assume that the errors were normally distributed, and that any trend that did occur was linear. The non-parametric approach is again simpler. The data are divided into two equal sets, the sequence being retained:

$$5.8\ 7.3\ 4.9\ 6.1\ 5.5$$
$$5.5\ 6.0\ 4.9\ 6.0\ 5.0$$

The result for the 6th day is then subtracted from that for the 1st day, that for the 7th day from that for the 2nd day etc. The signs of the differences between the pairs of values in the 5 columns are determined in this way to be $+ + 0 + +$. As usual, the zero value is ignored, leaving 4 results, all positive. The probability of obtaining 4 identical signs in four trials is clearly $2 \times (1/16) = 0.125$. (Note that a two-tailed test is again used, as the trend in the hormone level might be upwards or downwards.) The null hypothesis, that there is no trend in the results, can thus not be rejected at the 5% significance level.

The critical reader may find this result rather unsatisfactory. On inspection of the data it is hard to avoid the conclusion that, despite the result of the sign test, there is in fact a downward trend in the results. For example, the mean of the first three results is 6.00, the mean of the next four is 5.78, and the mean of the last three is 5.30. This apparent discrepancy is partly a result of the nature of hypothesis testing: we cannot reject the hypothesis that there is no trend in the results, but that does not mean that such a trend *must necessarily* be absent. But the example is also a reminder that the price paid for the extreme simplicity of the sign test is some loss of statistical power. The test does not utilize all the information offered by the data, so it is not surprising to find that it also provides less discriminating information. In later sections non-parametric methods that *do* use the magnitudes of the individual results as well as their signs will be discussed.

## 5.4 THE WALD–WOLFOWITZ RUNS TEST

In some instances we are interested not merely in whether observations generate positive or negative signs, but also in whether these signs occur in a random sequence. In Section 4.11, for example, we showed that if a straight line is a good fit to a set of calibration points, positive and negative residuals will occur more or less at random. By contrast, attempting to fit a straight line to a set of points that actually lie on a curve will yield non-random sequences of positive or negative signs. There might, for example, be a sequence of + signs, followed by a sequence of − signs, and then another sequence of + signs. Such sequences are technically known as **runs** − the word being used here in much the same way as when someone refers to 'a run of bad luck', or when a sportsman suffers 'a run of low scores'. In the curve-fitting case, it is clear that a non-random sequence of + and − signs will lead to a smaller number of runs than a random sequence. The Wald–Wolfowitz method tests whether the number of runs is small enough for the null hypothesis of a random distribution of signs to be rejected. The number of runs in the experimental data is compared with the numbers in Table A.8, which refers to the 5% significance level. The table is entered by using the appropriate values for $N$, the number of + signs, and $M$, the number of − signs. If the experimental number of runs is *smaller* than the tabulated value, then the null hypothesis can be rejected.

*Example.* Linear regression equations are used to fit a straight line to a set of 12 calibration points. The signs of the resulting residuals in order of increasing $x$ value are: + + + + − − − − − − + +. Comment on whether it would be better to attempt to fit a curve to the points.

Here it is clear that $N = M = 6$, and that the number of runs is 3. Table A.8 shows that, at the 5% significance level, the number of runs must be

<4 if the null hypothesis is to be rejected. Thus in this instance we can reject the null hypothesis, and conclude that the sequence of + and − signs is not a random one. The attempt to fit a straight line to the experimental points is therefore unsatisfactory, and a curvilinear regression plot is indicated instead.

There are two further points of interest in connection with the Wald-Wolfowitz test. First, it is worth noting that it can be used with any results that can be divided or converted into just two categories. Suppose, for example, that it is found that twelve successively used spectrometer light-sources last for 450, 420, 500, 405, 390, 370, 380, 395, 370, 370, 420 and 430 hours. The median lifetime, in this case the average of the sixth and seventh lamps when the data are arranged in ascending order, is 400 hours. If all those lamps with lifetimes less than the median are given a − sign, and all those with longer lifetimes are given a + sign, then the sequence becomes: + + + + − − − − − − + +. This is the same sequence as in the example above, and is shown in the same way to be significantly non-random. In this case, the significant variations in lifetime might be explained if the lamps came from different batches or different manufacturers.

It is also important to note that we may be concerned with unusually large numbers of short runs, as well as unusually small numbers of long runs. Thus, if 6 + and 6 − signs occurred in the order: + − + − + − + − + − + − we would strongly suspect a non-random sequence. Table A.8 shows that, with $N = M = 6$, a total of 11 or 12 runs indicates that the null hypothesis of a random order should be rejected, and some periodicity in the data suspected.

## 5.5 TESTS BASED ON THE RANGE OF THE RESULTS

In previous chapters, the standard deviation has been used as the most common measure of the dispersion or 'spread' of a set of results. In non-parametric statistics the **interquartile range** (see Section 5.1) is frequently used as a measure of dispersion. In addition, however, several very useful tests can be based on the whole range ($w$) of the sample, i.e. the difference between the largest and smallest values. It is evident that $w$ will be easier to calculate than the interquartile range, and the calculations for most of the tests described in this section can be made mentally. It must be emphasized, however, that although these tests are quick, *they are not* non-parametric: for example the calculations use the arithmetic mean, not the median, of the sample.

Simple range tests can be used in place of both the elementary applications of the $t$-test. In Section 3.2 the use of the $t$-test to compare an experimental mean with a known or standard value was described. In place of this test, the statistic $T_I$ can be calculated from the equation

$$T_I = |\bar{x} - V|/w \qquad (5.2)$$

where $\bar{x}$ is the mean, and $V$ the standard value.

*Example.* In a method for determining mercury by cold-vapour atomic-absorption spectrometry the following values were obtained for a standard material containing 38.9% mercury:

$$38.9, \ 37.4, \ 37.1\%$$

P.-K. Hou, O,-W. Lau and M.-C. Wong, *Analyst,* 1983, **108**, 64.

Is there any evidence of systematic error? Note that this example is the same as that in Section 3.2.

It is apparent that the mean, $\bar{x}$, in this case is 37.8%, and the range, $w$, is 1.8%. The value of $T_I$ is thus [Eq. (5.2)] $|37.8 - 38.9|/1.8 = 0.611$. This value is compared with that in Table A.9, for $n = 3$ and $P = 0.05$. The tabulated value is 1.304: since the experimental value is smaller, we cannot reject the null hypothesis, viz. that the experimental data could have come from a population with a mean value of 38.9% mercury. This result — that at the 5% significance level evidence of systematic error is lacking — is the same as that obtained in Section 3.2 by using the $t$-test, but the present calculation is obviously much simpler to perform. It should be noted that the $T_I$ test can also be used to estimate confidence limits. In this instance the confidence limits will be given by $x \pm T_I w$, i.e. by $37.8 \pm (1.304 \times 1.8) = 37.8 \pm 2.3\%$ — a range which, as expected, includes 38.9%.

A very similar test can be used in place of the $t$-test in the comparison of two mean values. If a set of data with mean $\bar{x}_1$ and range $w_1$ is to be compared with a second set with the same value of $n$ and with mean and range $\bar{x}_2$ and $w_2$ respectively, the statistic $T_d$ is calculated from:

$$T_d \ = \ 2 \, |\bar{x}_1 - \bar{x}_2|/(w_1 + w_2) \tag{5.3}$$

As usual, if the experimental value of $T_d$ is less than the tabulated value (Table A.10), then the null hypothesis, that the two sample means are equal, is retained. An example of this type of calculation is given at the end of the chapter. A further test, with an identical purpose, that has gained in popularity is **Lord's range test**. Here the test statistic, $L$, is given by:

$$L \ = \ |\bar{x}_1 - \bar{x}_2|/(w_1 + w_2) \tag{5.4}$$

i.e. it utilizes the sum of the ranges rather than the average range. These tests are useful in the comparison of small samples ($n$ may be as low as 2), but (like the $t$-test described in Section 3.3) assume that the parent population is normal, and that the spread is reasonably similar in the two samples. Again, therefore, their advantage is the simplicity of the calculations. A set of tabulated values for $L$ is given in Table A. 10: experimental values greater than the tabulated values allow the rejection of the null hypothesis.

Yet another simple range test can be used to replace the $F$-test in the comparison of the spread of two sets of results. Instead of comparing variances, the new statistic, $F_R$, compares ranges. That is, $F_R = w_1/w_2$ or $w_2/w_1$, whichever is greater than 1. The experimental $F_R$ value is then compared as usual with the tabulated values (Table A.11). If the experimental value is the higher, the null hypothesis, that the two samples come from populations of equal variances, is rejected. As in the real $F$-test, both one-tailed and two-tailed versions of this 'substitute $F$-test' are available, so Table A.11 must be used with care. An example of the use of this method is given in the exercises at the end of the chapter.

The range was used in another test described in Chapter 3 (Section 3.6) – the $Q$-test for outliers. It was shown that the experimental value of $Q$ [given by |(suspect value − nearest value)|/range] could be compared with a set of tabulated values (Table A.4) to establish whether a suspect measurement might be a candidate for rejection. It can be shown that a simpler but approximate form of this test dispenses with the need for the tables. An outlier can be rejected at the significance level $P = 0.05$ if $Q > \sqrt{(2/n)}$, and at the level $P = 0.01$ if $Q > \sqrt{(3/n)}$, where $n$ is the number of measurements, *including* the one that is suspect.

This test provides a reminder that range methods suffer from the disadvantage of being greatly affected by one or more outlying results. Truly non-parametric methods, in contrast, are generally immune to such results. Range tests also have a further disadvantage: they have less statistical power than the corresponding methods described in Chapter 3. Because range tests do not use all the available data, there will be cases where an $F$-test (for example) indicates that the appropriate null hypothesis can be rejected, but its simpler alternative, the substitute $F$-test, indicates that the null hypothesis should be retained. The popularity of range tests indicates that these are problems for which the simplicity of the calculations is more than adequate compensation.

## 5.6 THE WILCOXON SIGNED RANK TEST

Section 5.3 described the use of the sign test. Its value lies in the minimal assumptions it makes about the experimental data. The population from which the sample is taken is not assumed to be normal, or even to be symmetrical – the only prior information needed is the value of the median. A compensating disadvantage of the sign test is that it uses so little of the information provided. The only material point is whether an individual measurement is greater than or less than the median – the magnitude of this deviation is not used at all.

In many instances an analyst will have every reason to believe that his measurements will be *symmetrically* distributed but will not wish to make the assumption that they are normally distributed. This assumption of symmetrical data, and the consequence that the mean and the median of the population will be equal, allows more powerful significance tests to be developed. Important advances were made by Wilcoxon, and his signed rank test has several applications. Its mechanism is best illustrated by an example.

*Example.* The blood lead levels (in pg/ml) of seven children were found to be 104, 79, 98, 150, 87, 136 and 101. Could such data come from a population, assumed to be symmetrical, with median (mean) 95 pg/ml?

Compared with the reference value (95) the data have values of

$$9, -16, 3, 55, -8, 41, 6$$

These values are first arranged in order of magnitude without regard to sign, viz:

$$3, 6, 8, 9, 16, 41, 55$$

Their signs are then restored to them (in practice these last two steps would of course be combined):

$$3, 6, -8, 9, -16, 41, 55$$

The numbers are then **ranked**; in this process they keep their signs but are assigned numbers indicating their order (or rank), viz.

$$1, 2, -3, 4, -5, 6, 7$$

The positive ranks add up to 20, and the negative ones to 8. The *lower* of these two figures (8) is taken as the test statistic. The binomial theorem will give the probability of this number occurring. If the data come from a population with a median of 95 the sums of the negative and positive ranks would be expected to be approximately equal numerically; if the population median was very different from 95 the sums of the negative and positive ranks would be unequal. The probability of a particular sum occurring in practice is given by a set of tables (see Table A.12). In this test the null hypothesis is rejected if the tabulated value is *less than or equal to* the experimental value, i.e. the opposite of the situation encountered in most significance tests. In the present example, examination of the Table A.12 shows that, for $n = 7$, the test statistic must be less than or equal to 2 before the null hypothesis – that the data *do* come from a population with a median (mean) of 95 – can be rejected at a significance level of $P = 0.05$. In this example, the null hypothesis must clearly be retained. As usual, a two-tailed test is used, though there may be occasional cases where a one-tailed test is more appropriate.

An important advantage of the signed rank test is that it can also be used on paired data, because they can be transformed into the type of data given in the previous example. The signed rank method can thus be used as a non-parametric alternative to the paired *t*-test (Section 3.4).

*Example.* The following table gives the percentage concentration of zinc, determined by two different methods, for each of eight samples of health food.

| Sample | EDTA titration | Atomic spectrometry |
|--------|----------------|---------------------|
| 1      | 7.2            | 7.6                 |
| 2      | 6.1            | 6.8                 |
| 3      | 5.2            | 4.6                 |
| 4      | 5.9            | 5.7                 |
| 5      | 9.0            | 9.7                 |
| 6      | 8.5            | 8.7                 |
| 7      | 6.6            | 7.0                 |
| 8      | 4.4            | 4.7                 |

Is there any evidence for a systematic difference between the results of the two methods?

The approach to this type of problem is very simple. If there is no systematic difference between the two methods, then we would expect that the differences between the results for each sample, i.e. (titration result − spectrometry result), should be symmetrically distributed about zero. The signed differences are easily shown to be:

$$-0.4, -0.7, 0.6, 0.2, -0.7, -0.2, -0.4, -0.3$$

Arranging these values in numerical order without regard to sign, we have:

$$-0.2, 0.2, -0.3, -0.4, -0.4, 0.6, -0.7, -0.7$$

It is clear that the ranking of these values presents a difficulty, that of 'tied ranks'. There are two results with the numerical value 0.2, two with a numerical value of 0.4, and two with a numerical value of 0.7: how are the ranks to be calculated? This problem, in practice the only problem of any consequence encountered in ranking methods, is resolved by giving the tied values average ranks, with appropriate signs. Thus the ranking for the present data is:

$$-1.5, 1.5, -3, -4.5, -4.5, 6, -7.5, -7.5$$

In such cases, it is worth verifying that the ranking has been done correctly by calculating the sum of all the ranks without regard to sign. The sum for the numbers above is 36, which is the same as the sum of the first eight natural numbers, and therefore correct. The sum of the positive ranks is 7.5, and the sum of the negative ranks is 28.5. The test statistic is thus 7.5. Inspection of Table A.12 shows that, for $n = 8$, the test statistic has to be $\leqslant 3$ before the null hypothesis can be rejected at the significance level $P = 0.05$. In the present case, the null hypothesis must be retained − there is no evidence that the median (mean) of the differences is not zero, and hence no evidence for a systematic difference between the two analytical methods.

The signed rank test is seen from this example to be a simple and valuable method. Its principal limitation is that it cannot be applied to very small sets of data: for a two-tailed test at the significance level $P = 0.05$, $n$ must be at least 6.

## 5.7 THE WILCOXON RANK SUM TEST AND THE MANN–WHITNEY *U*-TEST

The signed rank test described in the previous section is valuable for the study of single sets of measurements, and for paired sets that can readily be reduced to single sets. In many instances, however, it is necessary to compare two independent samples that cannot be reduced to a single set of data. Such samples may contain different numbers of measurements. Several non-parametric tests to tackle such problems have been devised. The simplest to understand (though not, perhaps, the simplest to perform) is the Wilcoxon rank sum test. The operation of this test is most easily demonstrated by an example.

*Example.* A sample of photographic waste was analysed for silver by atomic-absorption spectrometry, five successive measurements giving values of 9.8, 10.2, 10.7, 9.5, and 10.5 µg/ml. After chemical treatment, the waste was analysed again by the same procedure, five successive measurements giving values of 7.7, 9.7, 8.0, 9.9 and 9.0 µg/ml. Is there any evidence that the treatment produced a significant reduction in the levels of silver?

The first step in the calculation is to rank *all* the data (i.e. from both sets of measurements). To distinguish the results obtained after the treatment we underline them:

$$\underline{7.7}, \underline{8.0}, \underline{9.0}, 9.5, \underline{9.7}, 9.8, \underline{9.9}, 10.2, 10.5, 10.7$$

The rankings, with those for the treated samples again underlined, are:

$$\underline{1}, \underline{2}, \underline{3}, 4, \underline{5}, 6, \underline{7}, 8, 9, 10$$

It is again apparent that, even when tied ranks arise and are treated as shown in the previous section, the sum of all the ranks must be $n(n + 1)/2$, in this case, 55.

If the measurements for the two sets of results were indistinguishable, we would expect that the underlined ranks and the other ranks should be mixed more or less at random in the list. That is, the sum of the underlined ranks and sum of the other ranks should be similar (if the numbers of measurements are the same in the two sets of results). In the present case the sum of the underlined ranks (treated samples) is 18 and the sum of the other ranks (untreated samples) is 37. These rank sums must now be converted into the test statistics $T_1$ and $T_2$, by using the equations:

$$T_1 = S_1 - n_1(n_1 + 1)/2 \tag{5.5}$$

$$T_2 = S_2 - n_2(n_2 + 1)/2 \tag{5.6}$$

where $S_1$ and $S_2$ are the sums of the two sets of ranks (18 and 37 in this example) and $n_1$ and $n_2$ are the corresponding numbers of observations. In this case $n_1 = n_2 = 5$, so $n_i(n_i + 1)/2 = 15$, and $T_1$ and $T_2$ are 3 and 22 respectively. The *lower* of these two values (3) is compared with the appropriate value in Table A.13. It is particularly important to note that in this example we can legitimately apply a *one-tailed* test. This is because the treatment method is specifically designed to *reduce* the silver level of the photographic waste: it may or may not do this successfully, but there is no realistic chance that it will actually increase the silver level! The table shows that, for $P = 0.05$ and $n_1 = n_2 = 5$, the lower $T$ value must not exceed 4 if the null hypothesis is to be rejected. It is thus possible to reject the null hypothesis — that the two samples came from the same population — and to conclude that the chemical treatment was effective in reducing the silver in the photographic waste.

The operation of the rank sum test is complicated by the calculation of the $T$ values from the rank sums. This additional step is essential, however, because the test is also valid when $n_1$ and $n_2$ are not equal. In such instances it is not reasonable to expect the sums of the two sets of ranks to be equal even if they are mixed more or less at random in the overall ranking, and Eqs. (5.5) and (5.6) make allowance for this effect. Because of this additional calculation step many workers prefer in practice to use the Mann–Whitney $U$-test. It will be seen that the $U$-text is a simpler but equally valid method of calculating the lower $T$ value defined above.

*Example.* Apply the Mann–Whitney $U$-test to the data of the previous example.

The procedure involves finding the number of results in one sample that exceed each of the values in the other sample. In the present case, we believe that the silver concentration for the treated solution should, if anything, be lower than that of the untreated solution. We thus expect to find that the number of cases in which a treated sample has a higher value than an untreated one should be small. Each of the values for the untreated sample is listed, and the number of instances where the values for the treated sample are greater are noted in each case.

| Untreated sample | Higher values in treated sample | Number of higher values |
|---|---|---|
| 9.8 | 9.9 | 1 |
| 10.2 | — | 0 |
| 10.7 | — | 0 |
| 9.5 | 9.7, 9.9 | 2 |
| 10.5 | — | 0 |

The total of the numbers in the third column is 3, the same $T$ value as that obtained in the rank sum test. It may be shown that the two methods always produce the same result, so Table A.13 is used for the Mann–Whitney $U$-test as well as for the rank sum test. It is clear that the $U$-test method is a much simpler way of arriving at the $T$ value. Indeed, when the numbers of measurements are small the calculation can be done mentally, a great advantage. If ties (identical values) occur in the $U$-test, each value is assigned a value of 0.5 in the count of $T$.

## 5.8 RANK CORRELATION

Ranking methods can also be applied to correlation problems. The Spearman rank correlation coefficient method to be described in this section is the oldest application of ranking methods in statistics, dating from 1904. Like other ranking methods, it is of particular advantage when one or both of the sets of observations under investigation can be expressed *only* in terms of a rank order rather than in quantitative units. Thus, in the following example, the possible correlation between the sulphur dioxide concentrations in a series of table wines and their taste quality is investigated. The taste quality of a wine is not easily expressed in quantitative terms, but it is relatively simple for a panel of wine-tasters to rank the wines in order of preference. Examples of other attributes that are easily ranked, but not easily quantified, include the condition of experimental animals, the quality of laboratory accommodation, and the efficiency of laboratory staff. It should also be remembered that if either or both of the sets of data under study should happen to be quantitative, then (in contrast to the methods described in Chapter 4) there is no need for them to be normally distributed. Like other non-parametric statistics, the Spearman rank correlation coefficient, $\rho$, is easy to determine and interpret. This is shown in the following example.

> *Example.* Seven different table wines are ranked in order of preference by a panel of experts; the best wine is ranked 1, the next best 2, and so on. The sulphur dioxide content (in parts per million) of each wine is then determined by flow-injection analysis with colorimetric detection. Use the following results to establish whether there is any relationship between wine quality and sulphur dioxide content.

| Wine: | A | B | C | D | E | F | G |
|---|---|---|---|---|---|---|---|
| Taste ranking: | 1 | 2 | 3 | 4 | 5 | 6 | 7 |
| $SO_2$ content: | 0.9 | 2.8 | 1.7 | 2.9 | 3.5 | 3.3 | 4.7 |

The first step in the calculation is to convert the sulphur dioxide concentrations from absolute values into rank values (if tied ranks occur they are treated as in the signed rank test described in section 5.6):

| Wine: | A | B | C | D | E | F | G |
|---|---|---|---|---|---|---|---|
| Taste ranking: | 1 | 2 | 3 | 4 | 5 | 6 | 7 |
| $SO_2$ ranking: | 1 | 3 | 2 | 4 | 6 | 5 | 7 |

The differences, $d_i$, between the two ranks are then calculated. They are clearly $0, -1, 1, 0, -1, 1, 0$. The correlation coefficient, $\rho$, is then given by:

$$\rho = 1 - \frac{6\sum_i d_i^2}{n(n^2 - 1)} \tag{5.7}$$

In this example, $\rho$ is $1 - (24/336)$, that is 0.929. Theory shows that, like the product–moment correlation coefficient, $\rho$ can vary between $-1$ and $+1$. When $n = 7$, $P$ must exceed 0.79 if the null hypothesis of no correlation is to be rejected at the significance level $P = 0.05$ (Table A.14). Here, we can conclude that there is a correlation between the sulphur dioxide content of the wines and their perceived quality. Bearing in mind the way the rankings were defined, there is strong evidence that higher sulphur dioxide levels produce less palatable wines!

Another rank correlation method, due to Kendall, was introduced in 1938. It claims to have some theoretical advantages over the Spearman method, but is much harder to calculate (especially when tied ranks occur) and is not so frequently used.

## 5.9 NON-PARAMETRIC REGRESSION METHODS

In the detailed discussion of linear regression methods in the previous chapter, the assumption of normally distributed $y$-direction errors was emphasized, and the complexity of some of the calculation methods was apparent. This complexity is largely overcome by the use of calculators or computers, but the interested reader can refer to the bibliography for sources of rapid approximate methods for fitting straight lines to data. There remains an interest in non-parametric methods for fitting a straight line to a set of points: of the several methods available, perhaps the simplest is Theil's 'incomplete' method, so called to distinguish it from another more complex procedure developed by the same author (the 'complete' method).

Theil's method assumes that a series of points $(x_1, y_1)$, $(x_2, y_2)$, etc. is being fitted by a line of the form $y = bx + a$. The first step in the calculation involves ranking the points in order of increasing $x$. If the number of points, $n$ is odd, the middle point, i.e. the median value of $x$, is deleted: the calculation always requires an even number of points. For any pair of points $(x_i, y_i)$, $(x_j, y_j)$ where $x_j > x_i$, the slope, $b_{ij}$, of the line joining the points can be calculated from:

$$b_{ij} = (y_j - y_i)/(x_j - x_i) \tag{5.8}$$

Slopes $b_{ij}$ are calculated for the pair of points $(x_1, y_1)$ and the point immediately after the median $x$-value, for $(x_2, y_2)$ and the second point after the median $x$-value, and so on until the slope is calculated for the line joining the point immediately before the median $x$ with the last point. Thus, if the original data contained 11 points, 5 slopes would be estimated (the median point having been omitted); 8 original points would give 4 slope estimates, and so on. These slope estimates are themselves arranged in ascending order and their median value, calculated as described in Section 5.2, is the estimated slope of the straight line. With this value of $b$, values $a_i$ for the intercept are estimated for each point with the aid of the equation $y = bx + a$. Again the estimates of $a$ are arranged in

ascending order and the median value chosen as the best estimate of the intercept of the line. The method is illustrated by the following example.

*Example.* The following results were obtained in a calibration experiment
for the absorptiometric determination of a metal chelate complex:

| Concentration, $\mu g/ml$: | 0 | 10 | 20 | 30 | 40 | 50 | 60 | 70 |
|---|---|---|---|---|---|---|---|---|
| Absorbance: | 0.04 | 0.23 | 0.39 | 0.59 | 0.84 | 0.86 | 1.24 | 1.42 |

Use Theil's method to estimate the slope and the intercept of the best
straight line through these points.

In this case the calculation is simplified by the occurrence of an even
number of observations, and by the fact that the $x$-values (i.e. the concentrations) occur at regular intervals and are already in ranking order. We thus
calculate slope estimates from 4 pairs of points:

$$b_{15} = (0.84 - 0.04)/40 = 0.0200$$
$$b_{26} = (0.86 - 0.23)/40 = 0.0158$$
$$b_{37} = (1.24 - 0.39)/40 = 0.0213$$
$$b_{48} = (1.42 - 0.59)/40 = 0.0208$$

We now arrange these slope estimates in order, obtaining 0.0158, 0.0200,
0.0208, 0.0213. The median estimate of the slope is thus the average of
0.0200 and 0.0208, i.e. 0.0204. We now use this value of $b$ to estimate the
intercept, $a$; the eight individual $a_i$ values are:

$$a_1 = 0.04 - (0.0204 \times 0) = +0.040$$
$$a_2 = 0.23 - (0.0204 \times 10) = +0.026$$
$$a_3 = 0.39 - (0.0204 \times 20) = -0.018$$
$$a_4 = 0.59 - (0.0204 \times 30) = -0.022$$
$$a_5 = 0.84 - (0.0204 \times 40) = +0.024$$
$$a_6 = 0.86 - (0.0204 \times 50) = -0.160$$
$$a_7 = 1.24 - (0.0204 \times 60) = +0.016$$
$$a_8 = 1.42 - (0.0204 \times 70) = -0.008$$

Arranging these intercept estimates in order, we have $-0.160$, $-0.022$,
$-0.018$, $-0.008$, $+0.016$, $+0.026$, $+0.040$. The median estimate is therefore $+0.004$. We have thus concluded that the best straight line is given by
$y = 0.0204x + 0.004$. The 'least-squares' line, calculated by the methods of
Chapter 4, is readily shown to be $y = 0.0195x + 0.019$. Figure 5.1 shows
that the two lines are very similar when plotted. However, the Theil method
has three distinct advantages: it does not assume that all the errors are in the
$y$-direction; it does not assume that either the $x$- or $y$-direction errors are
normally distributed; it is not affected by the presence of outlying results.
This last point is illustrated clearly by the point (50, 0.86) in the present
example. It has every appearance of being an outlier, but its value does not
affect the Theil calculation at all, since neither $b_{26}$ nor $a_6$ directly affects
the median estimates of the slope and intercept, respectively. In the least-
squares calculation, however, this outlying point carries as much weight as
the other points. This is reflected in the calculated results; the least-squares

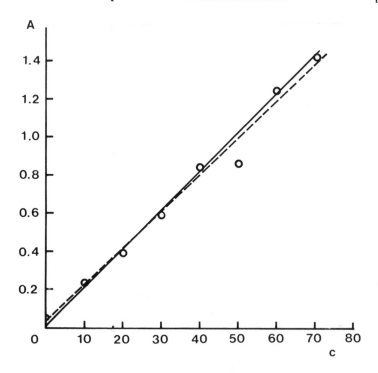

Fig. 5.1 − Straight-line calibration graph calculated by Theil's method (———),
and by the least-squares method of Chapter 4 (− − − − −).

line passes closer to the outlier than the non-parametric line does. Statistical
methods that are not greatly affected by outlying results are said to be
'robust' methods: their value in this and other cases is clear.

Unlike most non-parametric methods, Theil's method involves tedious
calculations, and a computer program would be a great advantage. It should be
noted that non-parametric methods for fitting curves are also available, but these
are beyond the scope of the present book.

## 5.10 THE KOLMOGOROV TEST FOR GOODNESS OF FIT

In Chapter 3 the common statistical problem of 'goodness of fit' was discussed.
This problem arises when it is required to test whether a sample of observations
might come from a particular distribution, such as a normal distribution. The
chi-squared test (Section 3.9) is very suitable for this purpose when the data are
presented as frequencies, though the test is not normally used for fewer than
about 50 observations, and is difficult to use with continuous data. In this
section the Kolmogorov method, which is well suited to testing goodness of fit

with continuous data, is described. Extensions of the method, which will not be described in detail, allow it to be extended to the comparison of two samples. These modified methods were first described by Smirnov, and the series of tests is often known as the Kolmogorov-Smirnov method.

The principle of the Kolmogorov approach is very simple. It involves comparing the cumulative frequency curve of the data to be tested with the cumulative frequency curve of the hypothesized distribution. The concept of the cumulative frequency curve, and its application in conjunction with normal probability paper, was discussed in Section 3.10. When the hypothetical and experimental curves have been drawn, the test statistic is obtained by finding the maximum vertical difference between them, and comparing this value in the usual way with a set of tabulated values. If the experimental data depart substantially from the expected distribution, the two curves will be expected to be widely separated over part of the cumulative frequency diagram: if, however, the data are closely in accord with the expected distribution, the two curves will never be very far apart. In practice, the Kolmogorov method has two common applications — testing for randomness, and testing for normality — and simple examples of these two applications will illustrate the operation of the procedure.

*Example.* The burette readings obtained by a laboratory worker over an extended period are analysed. In 50 readings the last recorded digits are found to be as follows.

| Digit:     | 0 | 1 | 2 | 3 | 4 | 5  | 6 | 7 | 8 | 9 |
|------------|---|---|---|---|---|----|---|---|---|---|
| Frequency: | 1 | 6 | 4 | 5 | 3 | 11 | 2 | 8 | 3 | 7 |

Are these readings consistent with a random use of digits 0-9? Note that the data in this example are identical to those of exercise 5 at the end of Chapter 3.

In this example the hypothetical distribution is a straightforward one: all the digits 0-9 are presumed to be equally likely, and no other results are possible. The hypothetical cumulative frequency plot is thus a regular step function as shown in Fig. 5.2. The cumulative frequencies in practice are clearly 0.02, 0.14, 0.22, 0.32, 0.38, 0.60, 0.64, 0.80, 0.86, 1.00 at $x = 0, 1, 2, \ldots 9$ respectively. The two plots show a maximum difference of 0.12 at $x = 4$. (In simple examples of this type it is unnecessary in practice to plot the cumulative frequencies — the maximum difference can be obtained by inspection.) Table A.15, which is suitable for a two-tailed test (the usual case), shows that at the significance level $P = 0.05$ for $n = 50$ the critical value is 0.188. So we cannot reject the null hypothesis that the digits are randomly distributed.

The reader will observe that this result is the same as that obtained by using the chi-squared method (Chapter 3, exercise 5). Closer inspection of the two tests shows that, in the chi-squared calculation, the experimental and tabulated

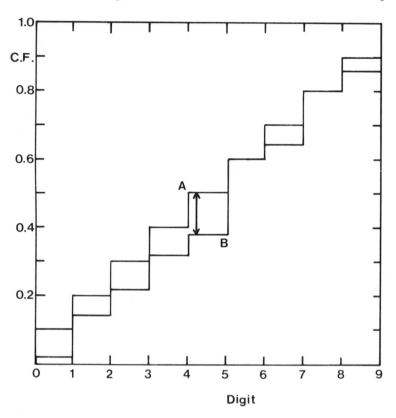

Fig. 5.2 – Kolmogorov's method used to test the randomness of digits 0–9. The maximum difference between the theoretical cumulative frequency plot (A) and the experimental plot (B), is shown by the arrow (←———→).

values are nearly the same, i.e. the titration digits only just fail to deviate significantly from a random distribution. In the Kolmogorov test, however, the result suggest that the data fail by a large margin to deviate from randomness. This is a reflection of the fact that the values in Table A.15 are based on the assumption that the distribution being tested is continuous. When the data are discrete, the results of the test are conservative, i.e. they tend to give too few significant results. There are, however, some advantages in the Kolmogorov test. In examples of the type just discussed, it is simpler to use, and is more reliable when $n$ is small. (The example above uses about the smallest set of data that can generally be studied by the chi-squared method, having a total of 50 observations, and an expected frequency of 5 for each number). The Kolmogorov method also gives non-parametric confidence limits for the true cumulative distribution function, by use of the data of Table A.15. This topic is covered more fully in detailed tests on non-parametric methods (see bibliography).

When the Kolmogorov method is used to test whether a distribution is normal, we must first transform the original data, which might have any values for their mean and standard deviation, into the **standard normal variable, $z$**. This is done by using the equation:

$$z = (x - \mu)/\sigma \qquad (5.9)$$

where the terms have their usual meanings. Statistical theory shows that $z$, calculated in this way, is itself normally distributed, and its cumulative distribution function is given in many collection of statistical tables. Equation (5.9) can be used in two ways. In some cases it may be required to test whether a set of data could come from a *particular* normal distribution, of *given* mean and standard deviation. In such a case, the experimental data are transformed directly by using Eq. (5.9), and the Kolmogorov test is performed. More often, it will simply be required to test whether the data might come from *any* normal distribution. In this instance, the mean and the standard deviation are estimated first, by the simple methods of Chapter 2; the data are next transformed by using Eq. (5.9); then the Kolmogorov method is applied. Both types of test are illustrated in the following example.

*Example.* Eight titrations were performed, with the results 25.13, 25.02, 25.11, 25.07, 25.03, 24.97, 25.14, and 25.09 ml. Could such results have come from (a) a normal population with mean 25.00 ml and standard deviation 0.05 ml, and (b) from any other normal population?

(a) Here, the first step is to transform the $x$-values into $z$-values, by using the relationship $z = (x - 25.00)/0.05$, obtained from Eq. (5.9). The eight results are thus transformed into 2.6, 0.4, 2.2, 1.4, 0.6, $-0.6$, 2.8, and 1.8. These $z$-values are arranged in order and plotted as a cumulative distribution function with a step height of 0.125 (i.e. 1/8). (Note that this is not the same calculation as that used in Section 3.10.) Comparison with the hypothetical function for $z$ indicates (Fig. 5.3) that the maximum difference is 0.545 at a $z$-value just below 1.4. To test this value, Table A.16 is used, Table A.15 being inappropriate once the data have been transformed into the standard form. Table A.16 shows that, for $n = 8$ and $P = 0.05$, the critical value is 0.288, so the null hypothesis can be rejected – the titration values probably do not come from a normal population with mean 25.00 ml and standard deviation 0.05 ml.

(b) In this case, we estimate the mean and the standard deviation [using Eqs. (2.1) and (2.2)] as 25.07 and 0.059 respectively, the latter result being correct to two significant figures. The $z$-values are now given by $z = (x - 25.07)/0.059$, i.e. by 1.02, $-0.85$, 0.68, 0, $-0.68$, $-1.69$, 1.19, 0.34. The cumulative frequency diagram for these values differs from the hypothetical curve by 0.125 at most (at several points). This difference is much smaller than

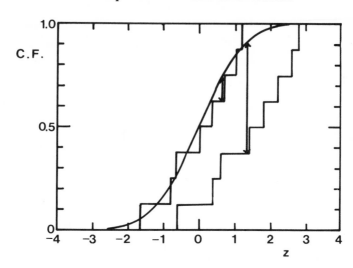

Fig. 5.3 – Kolmogorov's method used to test for the normal distribution. Maximum differences between the theoretical cumulative frequency curve and the two tested distributions are shown by the arrows (←———→).

the critical value of 0.288. We can thus accept that null hypothesis that the data come from a normal population with mean 25.07 and standard deviation 0.059.

## 5.11 CONCLUSIONS

The rapid and non-parametric tests described in this chapter are only a small fraction of the total available number of such methods. The examples given clearly exemplify their strengths and weaknesses. In most cases their speed and simplicity give them a distinct advantage over conventional methods, and the non-parametric tests do not involve the assumption of normality. They are ideally suited to preliminary examination of small samples, and to quick calculations made while the analyst is at the bench or on the shop floor. Their principal drawback is that they may be less discriminating than parametric tests when the latter can be validly applied. However, it seems certain that even the availability of microcomputers and other aids to calculation will not reduce interest in these very convenient tests.

## BIBLIOGRAPHY

W. J. Conover, *Practical Non-parametric Statistics,* Wiley, New York, 1971. Probably the best known general text on non-parametric methods.
W. W. Daniel, *Applied Nonparametric Statistics,* Houghton Mifflin, 1978. A very comprehensive text, covering a wide range of non-parametric methods in considerable detail; many examples.

R. Langley, *Practical Statistics*, Pan Books, 1968. A very readable book, with considerable emphasis on non-parametric tests.
P. Sprent, *Quick Statistics*, Penguin Books, 1981. An excellent introduction to non-parametric methods: clear, non-mathematical explanations, and many examples and exercises.

## EXERCISES

1. A titration was performed 4 times, with the results: 9.84, 9.91, 9.89, 10.20 ml. Calculate and comment on the median and the mean of these results.

2. The level of sulphur in batches of an aircraft fuel is claimed by the manufacturer to be symmetrically distributed with a median value of 0.10%. Successive batches are found to have sulphur concentrations of 0.09, 0.12, 0.10, 0.11, 0.08, 0.17, 0.12, 0.14 and 0.11%. Use the sign test, the signed rank test, and a range test to check the manufacturer's claim.

3. The concentrations (g/100 ml) of immunoglobulin G in the blood sera of ten donors is measured by radial immunodiffusion (r.i.d.) and by electroimmunodiffusion (e.i.d.), with the following results.

| Donor:        | 1   | 2   | 3   | 4   | 5   | 6   | 7   | 8   | 9   | 10  |
|---------------|-----|-----|-----|-----|-----|-----|-----|-----|-----|-----|
| R.i.d. result:| 1.3 | 1.5 | 0.7 | 0.9 | 1.0 | 1.1 | 0.8 | 1.8 | 0.4 | 1.3 |
| E.i.d. result:| 1.1 | 1.6 | 0.5 | 0.8 | 0.8 | 1.0 | 0.7 | 1.4 | 0.4 | 0.9 |

Are the results of the two methods significantly different?

4. In the development of a new method for the determination of blood alcohol levels, a single sample of blood is analysed 5 times, with the results 64.5, 66.0, 63.9, 65.1, and 64.0 mg/100 ml. The standard analysis method, applied to the same sample, gives the results 66.2, 65.8, 65.4, 66.3 and 65.6 mg/100 ml. Use statistical tests based on the ranges of these sets of results to test whether (a) the two methods give significantly different mean values, and (b) differ significantly in precision.

5. Ten carbon rods used successively in an electrothermal atomic-absorption spectrometer were found to last for 24, 26, 30, 21, 19, 17, 23, 22, 25 and 25 samples. Test the randomness of these rod lifetimes.

6. After each drinking three pints of beer, five volunteers were found to have blood alcohol levels of 104, 79, 88, 120 and 90 mg/100 ml. A further set of six volunteers drank three pints of lager each, and were found to have blood alcohol levels of 68, 86, 71, 79, 91 and 66 mg/100 ml. Use the rank sum test or the Mann–Whitney $U$-test to investigate the suggestion that drinking lager produces a lower blood alochol level than drinking the same amount of beer.

7. A university chemical laboratory contains seven atomic-absorption spectrometers (A–G). Surveys of the opinions of the research students and of the academic staff show that the students' order of preference for the instruments is B, G, A, D, C, F, E, and that the staff's order of preference is G, D, B, E, A, C, F. Are the opinions of the students and the staff correlated?

8. Use Theil's method to calculate the regression line for the data of exercise 1 in Chapter 4.

9. Taking the radial immunodiffusion data from exercise 3 above, use the Kolmogorov method to test the hypothesis that serum immunoglubulin G levels are normally distributed, with a mean of 1.0 g/100 ml and a standard deviation of 0.2 g/100 ml. Will any other normal distribution fit the data better?

# 6

# Experimental design, optimization and pattern recognition

## 6.1 INTRODUCTION

In Chapter 3 the idea of a **factor** was introduced. A factor is any aspect of the experimental conditions which affects the value obtained from an experiment. Two examples of such factors were discussed: Section 3.9 gave the example of the dependence of a fluorescence signal on the conditions under which a solution was stored. In this experiment the factor of interest was the storage conditions; it is called a **controlled factor** because it could be altered at will by the experimenter. In the other example, in which salt from different parts of a barrel was tested for purity, the factor of interest, i.e. the part of the barrel from which the salt was taken, was chosen at random. For that reason this factor is called an **uncontrolled factor**. Both these factors are **qualitative** since their possible 'values' cannot be arranged in numerical order. A factor for which the possible values *can* be arranged in numerical order is **quantitative** e.g. temperature. The different values which a factor takes are known as different **levels**.

As presented in Chapter 3, these experiments were intended as an introduction to the calculations involved in the analysis of variance (ANOVA). No mention was made of the experimental conditions, which might have introduced other factors that affected the results. In the fluorescence experiment such possible factors include the ambient temperature, whether the same fluorimeter was used for each measurement, and whether the measurements were made by the same person and on the same day. Any of these factors *might* have influenced the results to give the observed behaviour, thus invalidating the conclusion drawn about the effect of the storage conditions. Clearly, if the correct conclusions are to be drawn from an experiment, the various factors affecting the result must be identified and, if possible, controlled. The term **experimental design** is usually used to describe the stages of (1) identifying the factors which may affect the result of an experiment, (2) designing the experiment so that the

effects of uncontrolled factors are minimized, and (3) using statistical analysis
to separate the effects of the various factors involved.

## 6.2 RANDOMIZATION

One of the assumptions of ANOVA is that the uncontrolled variation is random.
However, in measurements made over a period of time, variation in an uncontrolled factor such as pressure, temperature, deterioration of apparatus etc. may
produce a trend in the results. As a result the errors due to uncontrolled variation
are no longer random, since successive errors are correlated. This can lead to a
systematic error in the results. Fortunately this problem is simply overcome by
using the technique of **randomization**. Suppose we wish to compare the effect
of a single factor at three different levels. In an experiment of this type the
different levels are often called **treatments**. The term treatment has its origin
in the development of ANOVA by Fisher for analysing agricultural experiments.
In such cases, different treatments might be, for example, the application of
different fertilizers. Let us denote three different treatments by A, B and C and
require that four replicate measurements are made for each treatment. Instead
of making the four measurements for treatment A, followed by the four for B,
then the four for C, we make the twelve measurements in a random order, decided
by using a table of random numbers. We assign each treatment a number for
each replication:

| A | B | C |
|---|---|---|
| 01 02 03 04 | 05 06 07 08 | 09 10 11 12 |

(Note that each number has the same number of digits.) We then enter a random
number table (see Table A.6) at an arbitrary point and read off pairs of digits,
discarding the pairs 00, 13-99 and repeats. Suppose this gives the sequence 02,
10, 04, 03, 11, 01, 12, 06, 08, 07, 09, 05; then the measurements would be
made at the different levels in the order A, C, A, A, C, A, C, B, B, B, C, B.
This random order of measurement ensures that the errors caused at level A
by uncontrolled factors are random, likewise for levels B and C.

## 6.3 BLOCKING

One disadvantage of complete randomization is that it fails to take advantage
of any natural subdivisions in the experimental material. Suppose, for example,
all the twelve measurements in the previous example could not be made on the
same day but had to be divided between four consecutive days. Using the same
order as before would give:

> Day 1   A C A
> Day 2   A C A
> Day 3   C B B
> Day 4   B C B

With this design the measurements using treatment A occur on the first two days whereas those using treatment B are made on the last two days. If it seemed that there was a difference between treatments A and B it would not be possible to tell whether this difference was genuine or if it was caused by the effect of using the two treatments on different pairs of days. A better design is one in which each treatment is used once on each day, with the order of the treatments randomized on each day. For example:

Day 1   A C B
Day 2   A B C
Day 3   C B A
Day 4   C A B

A group which contains one measurement for each treatment is known as a **block**, (again a term of agricultural origin signifying an area of a field) and the design is called a **randomized block** design. In this case the measurements on each day form a block.

Another advantage of a blocked experiment is that it is more sensitive than an unblocked one as a method of testing whether different treatments produce significantly different results. ANOVA — in this case two-way because each measurement is classified according to two factors, treatment and day — allows the separation of the three sources of variation: between-block variation, between-treatment variation, and random variation due to experimental error. In order to decide whether the treatments produce significantly different results, the between-treatment estimate of variance is compared with the variance estimated from the random errors. The sensitivity of the experiment depends on the size of the random variation: the smaller the random variation, the smaller the difference between the treatments which can be detected. In an unblocked experiment the random variation would be larger since it would include a contribution due to day-to-day variation, and consequently the sensitivity of the experiment would be reduced.

### 6.4 TWO-WAY ANOVA

In two-way ANOVA, each measurement $x_{ij}$ is classified according to two factors, as shown in Table 6.1. There are $N$ measurements divided between $c$ treatment levels and $r$ blocks (therefore $N = cr$). The column and row totals and the grand total, $T$, are also given, as these are needed in the calculation. The derivation of the formulae used will not be given in detail as they were for one-way ANOVA in Section 3.9. The principle of the method is the same, and references are given at the end of the chapter for those who wish to pursue the mathematics.

**Table 6.1** – General form of table for two-way ANOVA

| | 1 | 2 | ..... | $j$ | ..... | $c$ | Row total |
|---|---|---|---|---|---|---|---|
| | | | Treatment | | | | |
| Block 1 | $x_{11}$ | $x_{12}$ ..... | | $x_{1j}$..... | | $x_{1c}$ | $T_1.$ |
| Block 2 | $x_{21}$ | $x_{22}$ ..... | | $x_{2j}$..... | | $x_{2c}$ | $T_2.$ |
| . | . | . ..... | | ...... | | . | . |
| . | . | . ..... | | ...... | | . | . |
| Block i | $x_{i1}$ | $x_{i2}$ ..... | | $x_{ij}$ ..... | | $x_{ic}$ | $T_i.$ |
| . | . | . ..... | | ...... | | . | . |
| . | . | . ..... | | ...... | | . | . |
| Block r | $x_{r1}$ | $x_{r2}$ ..... | | $x_{rj}$ ..... | | $x_{rc}$ | $T_r.$ |
| Column total | $T._1$ | $T._2$ ..... | | $T._j$ ..... | | $T._c$ | $T$ = grand total |

The formulae for calculating the variation from the three different sources, i.e. between-treatment, between-block and experimental error, are given in Table 6.2.

**Table 6.2** – Formulae for two-way ANOVA

| Source of variation | Sum of squares | Degrees of freedom |
|---|---|---|
| Between-treatment | $\sum_{j=1}^{c} T_{.j}^2/r - T^2/N$ | $c-1$ |
| Between-block | $\sum_{i=1}^{r} T_i^2/c - T^2/N$ | $r-1$ |
| Residual | by subtraction | by subtraction |
| Total | $\sum x_{ij}^2 - T^2/N$ | $N-1$ |

The residual sum of squares and the number of degrees of freedom are found by subtracting the corresponding between-treatment and between-block values from the total values.

*Example*. In an experiment to compare the percentage efficiency of different chelating agents in extracting metal ions from aqueous solution the following results were obtained:

| | | Chelating agent | | |
|---|---|---|---|---|
| Day | A | B | C | D |
| 1 | 84 | 80 | 83 | 79 |
| 2 | 79 | 77 | 80 | 79 |
| 3 | 83 | 78 | 80 | 78 |

On each day a fresh solution of the metal ion (with the specified concentration) was prepared and the extraction performed with each of the chelating agents, taken in a random order.

In this experiment the use of different chelating agents is a controlled factor (see Section 3.8) since the chelating agents are chosen by the experimenter; the day on which the experiment is performed introduces uncontrolled variation caused both by changes in laboratory temperature, etc. and slight differences in the concentration of the metal ion solution i.e. the day is an uncontrolled factor. In Chapter 3 it was explained how ANOVA can be used either to test for a significant effect due to a controlled factor, or to estimate the variance of an uncontrolled factor. In this experiment, where there are both types of factor, two-way ANOVA can be used in both ways: (i) to test whether the different chelating agents have significantly different efficiencies, and (ii) to test whether the day-to-day variation is significantly greater than the variation due to the random error of measurement and, if it is, to estimate its variance.

As in one-way ANOVA, the calculations can be simplified by subtracting an arbitrary number from each measurement. The table below shows the measurements with 80 subtracted from each.

| Blocks | A | B | C | D | Row totals, $T_{i.}$ | $T_{i.}^2$ |
|---|---|---|---|---|---|---|
| | | Treatments | | | | |
| Day 1 | 4 | 0 | 3 | −1 | 6 | 36 |
| Day 2 | −1 | −3 | 0 | −1 | −5 | 25 |
| Day 3 | 3 | −2 | 0 | −2 | −1 | 1 |
| Column totals, $T_{.j}$ | 6 | −5 | 3 | −4 | 0 = Grand total, $T$ | |
| $T_{.j}^2$ | 36 | 25 | 9 | 16 | | |

$62 = \sum_i T_{i.}^2$

$$\sum_j T_{.j}^2 = 86$$

We have: $r = 3$, $c = 4$, $N = 12$, $\sum_i T_{i.}^2 = 62$, $\sum_j T_{.j}^2 = 86$, $T = 0$, $\sum x_{ij}^2 = 54$.

The calculation of the mean-square for each source of variation is shown in the next table.

| Source of variation | Sum of squares | Degrees of freedom | Mean square |
|---|---|---|---|
| Between-treatment | $86/3 - 0^2/12 = 28.6667$ | 3 | $28.6667/3 = 9.5556$ |
| Between-block | $62/4 - 0^2/12 = 15.5$ | 2 | $15.5/2 = 7.75$ |
| Residual | by subtraction = 9.8333 | 6 | $9.8333/6 = 1.6389$ |
| Total | $54 - 0^2/12 = 54.0$ | 11 | |

(Since the residual sum of squares involves calculating a difference, as many significant places as possible should be carried in order to avoid errors in this quantity.)

The reader may find it instructive to verify that the sums of squares *do* separate the treatment and block effects. This can be done by (say) increasing all the values in one block by a fixed number and recalculating the sums of squares. It should be found that although the total and between-block sums of squares are altered, the between-treatment and residual sums of squares are unchanged.

If there is no difference between the efficiencies or days then all three mean sums of squares should give an estimate of $\sigma_0^2$, the variance of the random variation due to experimental error (see Section 3.9). The $F$-test is used to see whether the estimates of variance differ significantly. Comparing the between-treatment mean square with the residual mean square gives:

$$F_{3,6} = 9.5556/1.6389 = 5.83$$

From Table A.2 the critical value of $F_{3,6}$ is 4.76 ($P = 0.05$) indicating that there is a significant difference between treatments, i.e. between the efficiency of different chelating agents, at the 5% level. Comparing the between-block (i.e. between-day) and residual mean squares gives:

$$F_{2,6} = 7.75/1.6389 = 4.76$$

The critical value of $F_{2,6}$ is 5.143 ($P = 0.05$) so there is no significant difference between days. Nevertheless the between-block mean square is considerably larger than the residual mean square and had the experiment been unblocked, so that these two effects were combined in the estimate of experimental error, the experiment would probably have been unable to detect whether different treatments gave significantly different results. If the difference between days *had* been significant it would indicate that other factors such as temperature, the preparation of the solution etc., were having an effect. It can be shown that the between-block mean square gives an estimate of $\sigma_0^2 + c\sigma_b^2$ where $\sigma_b^2$ is the variance of the random day-to-day variation. Since the residual mean square gives an estimate of $\sigma_0^2$, an estimate of $\sigma_b^2$ can be deduced.

This example illustrates clearly the benefits of considering the design of an experiment before it is performed. Given a blocked and an unblocked experiment with the same number of measurements in each, the former is more sensitive and yields more information.

The analysis above is based on the assumption that the effects, if any, of the two factors are additive. This point is discussed further in Section 6.7.

## 6.5 LATIN SQUARES

A uncontrolled factor which was not taken into account in the analysis of the previous section was the time of day at which the measurements were made. Systematic variation during the day due to the deterioration of the solution or rise in laboratory temperature could produce a trend in the results. It is possible to use an experimental design which allows the separation of this factor if there are equal numbers of blocks and treatments. Instead of, for example, the following randomized block design

|       |   |   |   |
|-------|---|---|---|
| Day 1 | A | C | B |
| Day 2 | C | A | B |
| Day 3 | C | B | A |

where the measurements using treatment C are concentrated at the beginning of the day and those using B are concentrated at the end of the day, the following design could be used:

| Day 1 | A | B | C |
|-------|---|---|---|
| Day 2 | C | A | B |
| Day 3 | B | C | A |

This design, in which each treatment appears once in each row and once in each column, is known as a **Latin square**; it allows the separation of the variation into between-treatment, between-day, between-time-of-day and random experimental error.

If the square is greater than 3 X 3 there is more than one possible design and a design should be chosen at random. The obvious disadvantage of a simple Latin square design is that it requires equal numbers of blocks and treatments. More complicated designs are possible which remove this constraint.

## 6.6 NESTED AND CROSS-CLASSIFIED DESIGNS

The experiment design given in Section 6.4 is known as **cross-classified** because a measurement is made for each possible combination of the different levels of the factors. Contrast it with the experimental design shown in Fig. 6.1, where a sample of a standard solution is sent to each of three laboratories and in each laboratory the sample is analysed by two different technicians. Such a design is said to be **nested** or **hierarchical** and it can be seen that not all possible combinations of experimental conditions are used, e.g. the technicians do not make measurements in laboratories other than their own. (One-way ANOVA can also be considered as a nested design.) Mixed (i.e. partly cross-classified, partly nested) designs are also possible — an example follows in the next section.

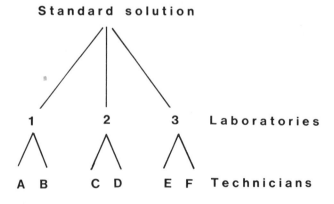

Fig. 6.1 – Nested or hierarchical design.

## 6.7 INTERACTION

A basic assumption of the analysis performed in Section 6.4 was that the effects of the two factors were **additive**. The meaning of this term is most easily explained by a numerical example. The simplest situation is that in which each factor has two levels: e.g. factor A has levels $a_1$ and $a_2$ and factor B has levels $b_1$ and $b_2$. Suppose, for example, the results for the combinations $a_1 b_1, a_1 b_2$ and $a_2 b_1$ are 10, 15 and 12 as shown in the table below.

**Table 6.3**

|  |  | Level of A | |
| --- | --- | --- | --- |
|  |  | $a_1$ | $a_2$ |
| Level of B | $b_1$ | 10 | 12 |
|  | $b_2$ | 15 | ? |

If these are the true values, i.e. there is no random variation, then the effect of changing factor B from level $b_1$ to level $b_2$ with A at $a_1$ is 5. If the effects are additive then the effect of changing B from level $b_1$ to level $b_2$ with A at $a_2$ should also be 5, so the final value in the table should be 17. Note that the effect of changing A from level $a_1$ to level $a_2$ is then 2, irrespective of the level of B. In general, if the results are as shown below:

|  |  | Level of A | |
| --- | --- | --- | --- |
|  |  | $a_1$ | $a_2$ |
| Level of B | $b_1$ | $r_1$ | $r_2$ |
|  | $b_2$ | $r_3$ | $r_4$ |

and they are free from random error, then $r_2 - r_1$ equals $r_4 - r_3$ if the effects are additive. If the effects are not additive, we say that there is **interaction** between A and B because a particular combination of A and B gives a higher value than expected. Figure 6.2 illustrates the effect of interaction graphically. On both graphs each result is plotted against the level of B, and points with the same level of A are joined. If the effects are additive then the lines will be parallel as in the left-hand diagram; if the effects interact the lines will not be parallel, as in the right-hand diagram. If both factors have several levels, a graph of this type is helpful in the interpretation of any interaction.

Unfortunately the experimental situation is confused by the presence of random errors. The reader may check that for Table 6.3, two-way ANOVA gives zero for the residual sum of squares, but if one of the values is altered this is no longer so. With this design of the experiment we cannot assess the extent to which a non-zero residual sum of squares is caused by random error, rather than by interaction between the two effects. In order to estimate the random error the measurements in each cell of the table must be repeated. The exact

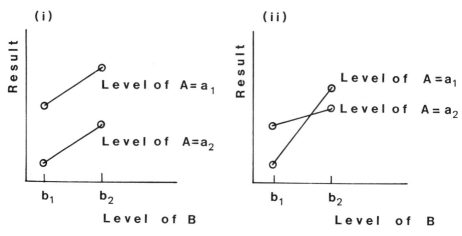

Fig. 6.2 – (i) Effects which are additive; (ii) effects which interact.

form of this repetition is important, as it must lead to an estimate of the random experimental error from all sources. Suppose, for example, a determination involves dissolving a given mass of a specimen in a specified volume of water and titrating against a standard solution. The repeated measurements should involve both the weighing and titration stages: if only the titration stage were repeated the random error in weighing would not be included in the estimate of experimental error. Further, if the same batch of standard solution and the same set of glassware is *not* used for *all* the determinations in the experiment, then the same batch of standard solution and the same set of glassware must *not* be used for the repeated measurements. Repeated measurements which are done so that they are subject to all the sources of random error in the experiment are called **replicates**. The method by which the random error and interaction sums of squares can then be separated is illustrated by the following example.

*Example.* In an experiment to investigate the viability of a solution as an absorbance standard, the value of the molar absorptivity, $\epsilon$, of solutions of three different concentrations was determined at four different wavelengths. Two replicate measurements were made for each combination of storage time and wavelength. The order in which the measurements were made was randomized. The results are shown below: for simplicity of calculation the measured values have been multiplied by 100.

**Table 6.4**

| Concentration (g/l.) $\backslash \lambda$(nm) | 240 | 270 | 300 | 350 |
|---|---|---|---|---|
| 0.02 | 94, 96 | 106, 108 | 48, 51 | 78, 81 |
| 0.06 | 93, 93 | 106, 105 | 47, 48 | 78, 78 |
| 0.10 | 93, 94 | 106, 107 | 49, 50 | 78, 79 |

As stated in Section 6.6, this is a mixed design: each cell of the table is cross-classified, with two replicate measurements in each cell.

The first stage of the calculation is to find the cell totals. This is done in the table below, which also includes other quantities needed in the calculation. As before, $T_i.$ denotes the total of the $i$th row, $T._j$ the total of the $j$th column, and $T$ the grand total.

**Table 6.5**

| Concentration (g/l.)\λ(nm) | 240 | 270 | 300 | 350 | $T_i.$ | $T_i^2$ |
|---|---|---|---|---|---|---|
| 0.02 | 190 | 214 | 99 | 159 | 662 | 438244 |
| 0.06 | 186 | 211 | 95 | 156 | 648 | 419904 |
| 0.10 | 187 | 213 | 99 | 157 | 656 | 430336 |

$$T._j \; 563 \quad 638 \quad 293 \quad 472 \qquad T = 1966 \qquad 1288484 = \sum_i T_{i.}^2$$

$$T._j^2 \; 316969 \; 407044 \; 85849 \; 222784$$

$$\sum_j T._j^2 = 1032646$$

As before, the between-row, between-column and total sums of squares are calculated. Each calculation requires the term $T^2/nrc$ (where $n$ is the number of replicate measurements in each cell, in this case 2, $r$ = number of rows and $c$ = number of columns). This term is sometimes called the **correction term, $C$**. We have:

$$C = T^2/nrc = 1966^2 /(2 \times 3 \times 4) = 161048.16$$

The sums of squares are now calculated:

Between-row sum of squares $= \sum_i T_i^2/nc - C$

$$= 1288484/(2 \times 4) - 161048.16$$
$$= 12.34$$

with $r - 1 = 2$ degrees of freedom

Between-column sum of squares $= \sum_j T._j^2/nr - C$

$$= 1032646/(2 \times 3) - 161048.16$$
$$= 11059.506$$

with $c - 1 = 3$ degrees of freedom.

Total sum of squares $= \sum x_{ijk}^2 - C$ where $x_{ijk}$ is the $k$th replicate in the $i$th row and $j$th column, i.e. $\sum x_{ijk}^2$ is the sum of the squares of the individual measurements in Table 6.4. Thus:

Total sum of squares $= 172138 - 161048.16$
$$= 11089.84$$

with $nrc - 1 = 23$ degrees of freedom.

The variation due to random error (usually called the **residual variation**) is estimated from the within-cell variation, i.e. the variation between replicates.

The residual sum of squares $= \Sigma x_{ijk}^2 - \Sigma T_{ij}^2/n$, where $T_{ij}$ is the total for the cell in the $i$th row and $j$th column, i.e. the sum of the replicate measurements in the $i$th row and $j$th column.

$$\text{Residual sum of squares} = \Sigma x_{ijk}^2 - \Sigma T_{ij}^2/n$$
$$= 172138 - 344244/2$$
$$= 16$$

with $rc(n-1) = 12$ degrees of freedom.

The interaction sum of squares and number of degrees of freedom can now be found by subtraction. The results of these calculations are summarized below.

| Source | Sum of squares | Degrees of freedom | Mean square |
|---|---|---|---|
| Between-row | 12.34 | 2 | 6.17 |
| Between-column | 11059.506 | 3 | 3686.502 |
| Interaction | 1.994 | 6 | 0.3323 |
| Residual | 16 | 12 | 1.3333 |
| Total | 11089.84 | 23 | |

Each source of variation is compared with the residual mean square to test whether it is significant.

(i) **Interaction.** This is obviously not significant since the interaction mean square is less than the residual mean square.

(ii) **Between-column** (i.e. between-wavelength). This is significant since we have:

$$F_{3,12} = 3686.502/1.3333 = 2765$$

(critical value of $F_{3,12} = 3.490, P = 0.05$)
In this case a significant result would be expected since absorbance is wavelength dependent.

(iii) **Between-row** (i.e. between-concentration). We have:

$$F_{2,12} = 6.17/1.3333 = 4.63$$

The critical value of $F_{2,12}$ is 3.885 ($P = 0.05$) indicating that the between-row variation is too great to be accounted for by random variation: the solution is not suitable as an absorbance standard. Figure 6.3 shows the molar absorptivity plotted against wavelength, with the values for the same concentration joined by straight lines. This illustrates the results of the analysis above in the following ways:

(i) the lines are parallel, indicating no interaction;
(ii) the lines are not horizontal, indicating that the molecular absorptivity varies with concentration;

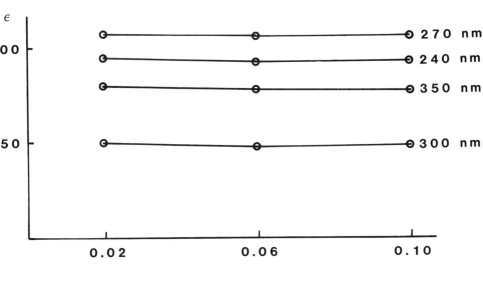

Fig. 6.3 – Relationships in the two-way ANOVA example of Section 6.7.

(iii) the lines are at different heights on the graph, indicating that the molar absorptivity is wavelength-dependent.

The formulae used in the calculation above are summarised in the table below.

**Table 6.6** – Formulae for two-way ANOVA with interaction

| Source of variation | Sum of squares | Degrees of freedom |
|---|---|---|
| Between-row | $\sum_i T_{i.}^2/nc - C$ | $r - 1$ |
| Between-column | $\sum_j T_{.j}^2/nr - C$ | $c - 1$ |
| Interaction | by subtraction | by subtraction |
| Residual | $\sum x_{ijk}^2 - \sum T_{ij}^2/n$ | $rc(n - 1)$ |
| Total | $\sum x_{ijk}^2 - C$ | $rcn - 1$ |

In this experiment both factors, i.e. the wavelength and the concentration of the solution, are controlled factors. In analytical chemistry an important application of ANOVA is the investigation of two or more controlled factors and their interactions in optimization experiments. This is discussed in Section 6.9.

Another important application of ANOVA is in collaborative investigations of precision between laboratories. An example of such an investigation would be one in which several different samples were sent to a number of laboratories with instructions to perform two replicate analyses on each sample. The first test would be whether there is any interaction between laboratory and sample, i.e. whether some laboratories obtained unexpectedly high or low results for some samples. If *no* interaction is found, a test is made of whether the laboratories obtained significantly different results, i.e. whether there is any systematic difference between laboratories. If there *is* a significant interaction between laboratory and sample, then testing for a significant difference between laboratories has little relevance.

Two-way ANOVA with interaction can also be applied when only one or neither of the factors is controlled. The calculation of the sums of squares is performed in the same way but the mean squares are compared slightly differently.

For two-way ANOVA to be valid the following conditions must be fulfilled (see Section 3.10):

(i) the random error is the same for all combinations of the levels of the factors;

(ii) the random errors are approximately normally distributed.

## 6.8 FACTORIAL VERSUS ONE-AT-A-TIME DESIGN

An experiment such as the one in the previous example, where the response variable (i.e. the molar absorptivity) is measured for all possible combinations of the chosen factor levels, is known as a **complete factorial design**. The reader may have noticed that this design of experiment is the antithesis of the classical approach in which the response is investigated for each factor in turn while all the other factors are held at a constant level. There are two cogent reasons for using a factorial rather than a classical design in experiments which test whether the response depends on factor level:

(i) the factorial experiment detects and estimates any interaction, which the one-at-a-time experiment cannot do;

(ii) if the effects of the factors are additive, then the factorial design needs fewer measurements than the classical approach in order to give the same precision. This can be seen by turning again to the molar absorptivity experiment. There, all 24 measurements were used to estimate the effect of varying the wavelength and the *same* 24 were used to estimate the effect of varying the concentration. In a one-at-a-time experiment, first the concentration would have been fixed and, to obtain the same precision for the effect of varying wavelength, 6 measurements would have been made at each wavelength, i.e. 24 in all. Then the wavelength would have been fixed and another 24 measurements made at different concentrations, making a total of 48 measurements. In general, for $k$ factors, a classical approach involves $k$ times as many experiments as a factorial one with the same precision.

## 6.9 FACTORIAL DESIGN AND OPTIMIZATION

Since analytical techniques are frequently concerned with detecting minute traces of analyte, it is usually important that the levels of the factors on which the response depends are chosen so as to maximize this response. The process of finding these optimum factor levels is known as **optimization**. The first step is to determine which factors and which interactions between them are important in affecting the response. This can be done by using a factorial experiment with each factor at two levels, usually known as 'low' and 'high'. In the case of a quantitative variable the terms 'low' and 'high' have their usual meaning. The exact choice of levels is determined principally by the experience and knowledge of the experimenter and the physical constraints of the system, e.g. if water is used as a solvent only temperatures in the range 0-100°C are practicable. Some problems attendant on the choice of levels are discussed below. For a qualitative variable, 'high' and 'low' refer to different conditions, e.g. the presence or absence of a catalyst, mechanical or magnetic stirring, a sample in powdered or granular form, etc. Since we have already considered two-factor experiments in some detail we will turn to one with three factors: A, B and C. This means that there are $2 \times 2 \times 2 = 8$ possible combinations of factor levels, as shown in the table below. A plus sign denotes that the factor is at the high level and a minus sign that it is at the low level. The last column gives a notation often used to describe the combinations, where the presence of the appropriate lower case letter indicates that the factor is at the high level and its absence that the factor is at the low level. The number 1 is used to indicate that all factors are at the low level.

| Response | A | B | C | Combination |
|----------|---|---|---|-------------|
| y | — | — | — | 1 |
| y | + | — | — | a |
| y | — | + | — | b |
| y | — | — | + | c |
| y | — | + | + | bc |
| y | + | — | + | ac |
| y | + | + | — | ab |
| y | + | + | + | abc |

The method by which the effects of the factors and their interactions are estimated is illustrated by the following example.

*Example.* In a high-performance liquid chromatography experiment, the dependence of the retention parameter, $k'$, on three factors was investigated. The factors were pH (factor P), the concentration of a counter-ion (factor T) and the concentration of the organic solvent in the mobile phase (factor C). Two levels were used for each factor and two replicate measurements made for each combination. The measurements were randomized. The table below gives the average value for each pair of replicates.

| Combination of factor levels | $k'$ |
|:---:|:---:|
| 1 | 4.7 |
| p | 9.9 |
| t | 7.0 |
| c | 2.7 |
| pt | 15.0 |
| pc | 5.3 |
| tc | 3.2 |
| ptc | 6.0 |

### (a) Effect of individual factors

The effect of changing the level of P can be found from the average difference in response when P changes from low to high level with the levels of C and T fixed. There are four pairs of responses which give an estimate of the effect of the level of P, as shown in the table below.

| Level of C | Level of T | Level of P + | Level of P − | Difference |
|:---:|:---:|:---:|:---:|:---:|
| − | − | 9.9 | 4.7 | 5.2 |
| + | − | 5.3 | 2.7 | 2.6 |
| − | + | 15.0 | 7.0 | 8.0 |
| + | + | 6.0 | 3.2 | 2.8 |
| | | | Total = | 18.6 |

Average effect of altering the level of P $= 18.6/4 = 4.65$

The average effect of altering the levels of T and C can be found in a similar way. It is left as an exercise for the reader to show that they are:

average effect of altering the level of C $= -4.85$
average effect of altering the level of T $= 2.15$.

### (b) Interaction between two factors

Consider now the two factors P and T. If there is *no* interaction between them, then the change in response between the two levels of P should be independent of the level of T. The first two figures in the last column of the table above give the change in response when P changes from high to low level with T at low level. Their average is $(5.2 + 2.6)/2 = 3.9$. The last two figures in the same column give the effect of changing P when T is at high level. Their average is $(8.0 + 2.8)/2 = 5.4$. If there is no interaction and no random error (see Section 6.7) these estimates of the effect of changing the level of P should be equal. The convention is to take half their difference as a measure of the interaction:

$$\text{effect of TP interaction} = (5.4 - 3.9)/2 = 0.75$$

It is important to realize that this quantity estimates the degree to which the effects of P and T are not additive. It could equally well have been calculated by considering how far the change in response for the two levels of T is independent of the level of P.

The other interactions are calculated in a similar fashion. Again the reader may check that:

effect of CP interaction $= -1.95$

effect of CT interaction $= -1.55$

## (c) Interaction between three factors

The PT interaction calculated above can be split into two parts according to the level of C. With C at low level the estimate of interaction would be $(8.0 - 5.2)/2 = 1.4$ and with C at high level it would be $(2.8 - 2.6)/2 = 0.1$. If there is no interaction between all three factors and no random error, these estimates of the PT interaction should be equal. The three-factor interaction is estimated by half their difference $[= (0.1 - 1.4)/2 = -0.65]$. The three-factor interaction measures the extent to which the effect of the PT interaction and the effect of C are not additive: it could equally well be calculated by considering the difference between the PC estimates of interaction for low and high levels of T or the difference between the TC estimates of interaction for low and high levels of P.

These results are summarised in the table below.

|  | Effect |
|---|---|
| Single factor (main effect) | |
| P | 4.65 |
| T | 2.15 |
| C | −4.85 |
| Two-factor interactions | |
| TP | 0.75 |
| CT | −1.55 |
| CP | −1.95 |
| Three-factor interactions | |
| PTC | −0.65 |

The calculations have been presented in some detail in order to make the principles clear. An algorithm due to Yates (see references) simplifies the calculation.

In order to test which effects, if any, are significant, ANOVA may be used (provided that there is homogeneity of variance). It can be shown that in a *two*-level experiment, like this one, the required sums of squares can be calculated from the estimated effects by using

$$\text{Sum of squares} = N \times (\text{estimated effect})^2/4$$

where $N$ is the total number of measurements, including replicates. In this case $N$ is 16 since two replicate measurements were made for each combination of factor levels. The calculated sums of squares are given below:

| Factor(s) | Sum of squares |
|---|---|
| P | 86.49 |
| T | 18.49 |
| C | 94.09 |
| PT | 2.25 |
| TC | 9.61 |
| PC | 15.21 |
| PTC | 1.69 |

It can be shown that each sum of squares has one degree of freedom and since the mean square is given by:

mean square = sum of squares/number of degrees of freedom

each mean square is simply the corresponding sum of squares. To test for the significance of an effect, the mean square is compared with the error (residual) mean square. This is calculated from the individual measurements by the method described in the molar absorptivity example in Section 6.7. Suppose that the calculated residual mean square is found to be 0.012 (8 degrees of freedom). Testing for significance, starting with the highest order interaction, we have for the PTC interaction:

$$F_{1,8} = 1.69/0.012 = 141$$

which is obviously significant. If there is interaction between all three factors there is no point in testing whether the factors taken in pairs or singly are significant, since all factors will have to be considered in any optimization process. A single factor should only be tested for significance if it does not interact with other factors.

One problem with a complete factorial experiment such as this is that the number of experiments required rises rapidly with the number of factors: for $k$ factors at 2 levels with 2 replicates for each combination of levels, $2^{k+1}$ experiments are necessary, e.g. for 5 factors, 64 experiments. When there are more than three factors some economy is possible by assuming that 3-way and higher order interactions are negligible. The sums of squares corresponding to these interactions can then be combined to give an estimate of the residual sum of squares and replicate measurements are no longer necessary. The rationale for this approach is that higher order effects are usually much smaller than main effects and two-factor interaction effects. If higher order interactions *can* be assumed negligible, a suitable fraction of all possible combinations of factor levels is sufficient to provide an estimate of the main and two-factor interaction effects. Such an experimental design is called a **fractional factorial design.**

One problem in using a factorial design to determine which factors have a significant effect on the response is that, for factors which are continuous variables, the effect depends on the high and low levels used. If the high and low levels are too close together, the effect of the corresponding factor may be found not significant despite the fact that over the whole possible range of factor levels the effect of this factor is *not* negligible. On the other hand, if the levels are further apart they may fall on either side of a maximum, and still give a difference in response which is not significant.

### 6.10 INADEQUACY OF THE ONE-AT-A-TIME METHOD OF OPTIMIZATION

Section 6.8 demonstrated the superiority of a factorial design over a one-at-a-time experiment as a means of estimating whether a factor had a significant effect. The one-at-a-time method is also unsatisfactory for achieving optimum conditions if the response depends on two or more factors which interact (see Section 1.6). This can be demonstrated graphically when the response depends on two factors which are continuous variables. In this case the relationship between the response and the levels of the two factors can be represented by a surface in three dimensions as shown in Fig. 6.4. This surface is known as a **response surface.** It is more conveniently shown as a contour diagram, see Fig. 6.5, where the response on each contour is constant. The form of the contour lines is, of course, unknown to the experimenter, who wishes to determine the optimum levels, $x_0$ and $y_0$ for the factors X and Y respectively. Adopting a one-at-a-time approach with the initial level of X fixed at $x_1$, say, and varying the level of Y, would give a maximum response at the point A, when the level of Y is $y_1$. Next, holding the level of Y at $y_1$ and varying the level of X would give a maximum at B. Obviously this is not the true maximum and the position attained depends on the initial level chosen for $x_1$. The process could be repeated by varying the levels of X and Y alternately but this is an inefficient method of approaching $(x_0, y_0)$. Furthermore, in some cases it would not lead to the true maximum, as illustrated by Fig. 6.6 where, although C is not the true maximum, the response falls on either side of it in both the X and Y directions. A one-at-a-time method arriving at this point would therefore conclude that it represented the maximum response. Similar arguments apply to optimization of more than two factors, but the response surface cannot be easily visualized in these cases. For this reason the discussion of optimization methods will largely be confined to two factors, although the methods can be generalized for any number of factors.

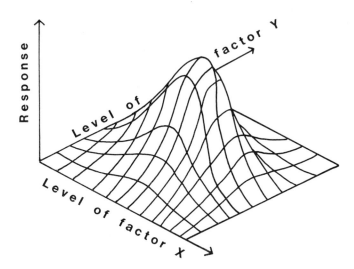

Fig. 6.4 – A response surface for two factors.

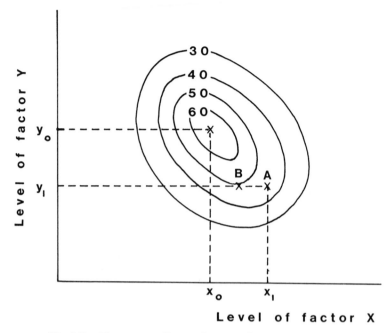

Fig. 6.5 – The contour diagram for a two-factor response surface.

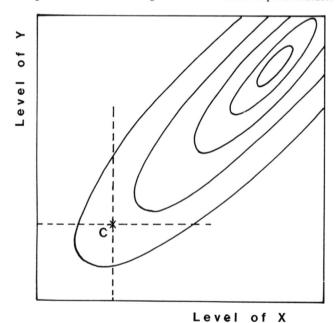

Fig. 6.6 – Contour diagram: a situation in which the one-at-a-time method fails to locate the maximum.

## 6.11 METHOD OF STEEPEST ASCENT

The process of optimization can be visualized in terms of a man standing on a hill in a thick fog with the task of finding the summit! In these circumstances an obvious approach is to walk in the direction in which the hill is steepest. This is the basis of the **method of steepest ascent**. Figure 6.7 shows two possible contour maps. A little consideration shows that the direction of steepest ascent at any point is at right angles to the contour lines at that point, as indicated by the arrows. When the contour lines are circular this will be towards the summit but when the contour lines are elliptical it will not. The shape of the contour lines

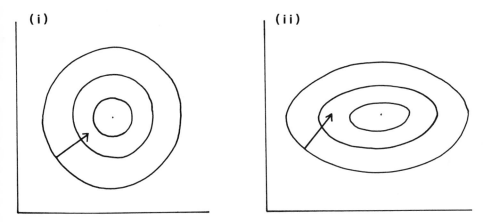

Fig. 6.7 – Contour diagrams: the arrow in each diagram indicates the path of steepest ascent. In (i) it goes close to the maximum but in (ii) it does not.

depends on the scales chosen for the axes: the best results are obtained from the method if the axes are scaled so that a unit change in either direction gives a roughly equal change in response. The first step is to perform a factorial experiment with each factor at two levels. The levels are chosen so that the design forms a square as shown in Fig. 6.8. Suppose, for example, that the experiment is an enzyme-catalysed reaction in which the reaction rate, which in this case is the response, is to be maximized with respect to pH (X) and temperature (Y). The table below gives the results (reaction rate, measured in arbitrary units) of the initial factorial experiment.

|        | pH 6.8 | pH 7.0 |
|--------|--------|--------|
| 20°C   | 30     | 35     |
| 25°C   | 34     | 39     |

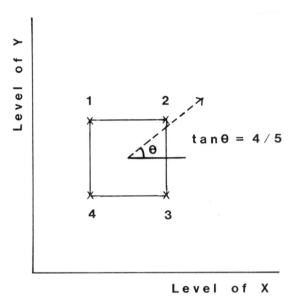

Fig. 6.8 – A 2 × 2 factorial design to determine the direction of steepest ascent, indicated by the broken line.

The effects of the two factors can be separated as described in Section 6.9. Rewriting the table above, with the notation of that section, gives:

| Rate of reaction | Combination of levels |
|---|---|
| 30 | 1 |
| 35 | x |
| 34 | y |
| 39 | xy |

Average effect of change in level of $X = \left\{(35 - 30) + (39 - 34)\right\}/2 = 5$
Average effect of change in level of $Y = \left\{(34 - 30) + (39 - 35)\right\}/2 = 4$

The effects of X and Y indicate that in Fig. 6.8 we should seek for the maximum response to the right and above the original region. Since the change in the X direction is greater than that in the Y direction the distance to be moved in the X direction should also be greater. To be more exact, the distances to be moved in the X and Y directions respectively should be in the ratio 5:4, i.e. in the direction indicated by the broken line in Fig. 6.8.

The next step in the optimization is to perform further experiments in this direction, as indicated by the broken line in Fig. 6.9, at (say) the points numbered 5, 6 and 7. This would indicate point 6 as a rough position for the maximum in this direction. Another factorial experiment is then performed in the region of point 6 to determine the new direction of steepest ascent.

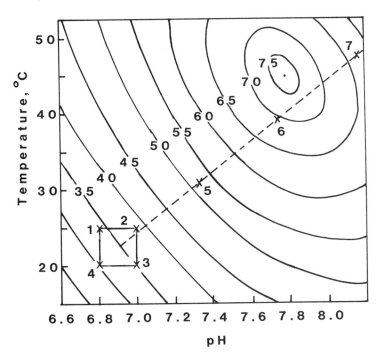

Fig. 6.9 – Contour diagram: the initial direction of steepest ascent is shown by
the broken line. Further experiments are done at points 5, 6 and 7.

This method gives satisfactory progress towards the maximum, provided
that over the region of the factorial design the contours are approximately straight.
This is equivalent to the response surface being planar, which in turn implies
that X and Y are independent. The indepence of X and Y can be checked by
replication and significance testing as described in Section 6.9. If X and Y are
*not* independent the response surface is curved. Its form can be determined
approximately by means of a more complicated factorial experiment with three
levels. The contours of this curved surface indicate the approximate position of
the maximum: if necessary, further factorial experiments can be performed to
locate the position more accurately.

As with all factorial designs the number of experiments required rises
sharply with the number of factors involved. This method of optimization also
involves complex calculations. The next section describes another method of
optimization which is conceptually simple and requires relatively little calculation.

## 6.12 SIMPLEX OPTIMIZATION

Simplex optimization may be applied when all the factors are continuous variables. (The reader should be warned that the choice of name is unfortunate since the method is different from the simplex methods used in linear programming.) **A simplex** is a geometrical figure which has $n + 1$ vertices when a response is being optimized with respect to $n$ factors. For example, for two factors the simplex will be a triangle. The method of optimization is illustrated by Fig. 6.10. The initial simplex is defined by the points labelled 1, 2 and 3. In the first experiment the response is measured at each of the combinations of factor levels given by the vertices of the simplex. The worst response would be found at point 3 and it would be logical to conclude that a better response would be obtained at a point which is the reflection of 3 with respect to the line joining 1 and 2, i.e. at 4. The points 1, 2 and 4 form a new simplex and the response is measured for the combination of factor levels given by 4. Comparing the responses for the points 1, 2 and 4 will show that 1 gives the worst response. The reflection procedure is repeated to give the simplex defined by 2, 4 and 5. The continuation of this process is shown in Fig. 6.10. It can be seen that no further progress is possible beyond the stage shown, since points 6 and 8 both give a worse response than 5 and 7.

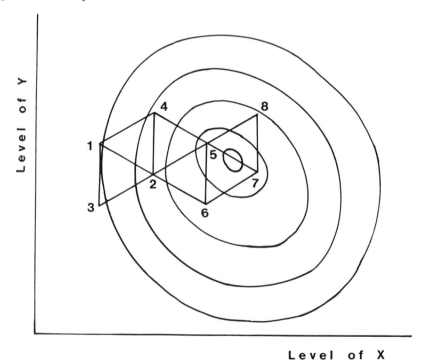

Fig. 6.10 – Simplex optimization.

In order to improve the performance of the simplex method various modifications have been (and still are being) proposed. The advance towards the maximum can be accelerated by using a simplex which can vary in size according to how the response for the new vertex in a simplex compares with the other vertices. Initially the simplex is large to give rapid progress towards the maximum; near the maximum it contracts to allow the maximum to be found accurately and to avoid the situation shown in Fig. 6.10.

The position of the new vertex of a simplex can be found by calculation rather than drawing — this is essential when there are more than two factors. The calculation is most easily set out in a table, as shown below. In this example there are 5 factors and hence the simplex has 6 vertices. The response for vertex 4 is lowest and so this vertex is to be replaced.

|  | | | Factors | | | |
|---|---|---|---|---|---|---|
|  | A | B | C | D | E | Response |
| Vertex 1 | 1.0 | 3.0 | 2.0 | 6.0 | 5.0 | 7 |
| Vertex 2 | 6.0 | 4.3 | 9.5 | 6.9 | 6.0 | 8 |
| Vertex 3 | 2.5 | 11.5 | 9.5 | 6.9 | 6.0 | 10 |
| Vertex 4 (rejected) | 2.5 | 4.3 | 3.5 | 6.9 | 6.0 | 6 |
| Vertex 5 | 2.5 | 4.3 | 9.5 | 9.7 | 6.0 | 11 |
| Vertex 6 | 2.5 | 4.3 | 9.5 | 6.9 | 9.6 | 9 |
| (i) Sum (excluding vertex 4) | 14.50 | 27.40 | 40.00 | 36.40 | 32.60 | |
| (ii) Sum/$n$ (excluding vertex 4) | 2.90 | 5.48 | 8.00 | 7.28 | 6.52 | |
| (iii) Rejected vertex (i.e. 4) | 2.50 | 4.30 | 3.50 | 6.90 | 6.00 | |
| (iv) Displacement = (ii) − (iii) | 0.40 | 1.18 | 4.50 | 0.38 | 0.52 | |
| (v) Vertex 7 = (ii) + (iv) | 3.30 | 6.66 | 12.50 | 7.66 | 7.04 | |

The coordinates of the centroid of the vertices which are to be *retained* are found by (i) summing the coordinates for the retained vertices and (ii) dividing by the number of factors, $n$. The displacement of the new point from the centroid is given by (iv) = (ii) − (iii), and the coordinates of the new vertex, vertex 7, by (v) = (ii) + (iv). If the simplex is to be varied in size then the values in row (iv) are multiplied by a suitable scaling factor.

An obvious question in using the simplex method is the choice of the initial simplex. If this is taken as a *regular* figure in $n$ dimensions, then the positions taken by the vertices in order to produce such a figure will depend on the scales used for the axes. As with the method of steepest ascent these scales should be chosen so that unit change in each factor gives roughly the same change in response. If there is insufficient information to achieve this, the difference between the highest and lowest feasible values of each factor can be represented by the same distance. The choice of the size of the initial simplex is discussed in a paper by Yarbro and Deming (see bibliography), who show that the size of the initial simplex is not critical if it can be expanded or contracted as the method proceeds. Yarbro and Deming quote an algorithm which can be used to calculate the initial positions of the vertices; one vertex is normally positioned at the currently accepted levels of the factors.

It can be seen that in contrast to a factorial design the number of experiments required in the simplex method does not increase rapidly with the number of factors. For this reason all factors which may be thought to have a bearing on the result should be included in the optimization, since doing so will not greatly increase the number of experiments required to define the optimum.

Once an optimum has been found, the effect on the response when one factor is varied while the others are held at their optimum levels can be investigated for each factor in turn. This procedure can be used to check the optimization. It also indicates how important deviations from the optimum level are for each factor: the sharper the response peak in the region of the optimum the more critical any variation in the factor level.

Simplex optimization has been used with success in many areas of analytical chemistry, e.g. atomic-absorption spectroscopy, gas chromatography, colorimetric methods of analysis, plasma spectrometry, and centrifugal analysers in clinical chemistry. When an instrument is interfaced with a microcomputer, the results of simplex optimization can be used to initiate automatic improvements in the instrument variables.

For response surfaces with more than one peak, such as that shown in Fig. 6.11, both types of optimization method described may locate a local optimum

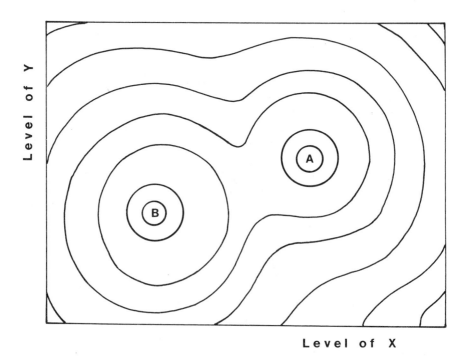

Fig. 6.11 – Contour diagram showing two localized maxima.

such as A rather than the true optimum B. Starting in a different region and seeing whether the same optimum conditions are obtained provides some means of checking this point. Optimization methods provide information about only *some* parts of the response surface, since their object is to maximize the response with the least effort. If a complete picture of the response surface is required, more elaborate experiments and analysis are needed.

## 6.13 PATTERN RECOGNITION

The advance of automation has made possible the rapid collection of large amounts of data, and the development of microcomputers has made sophisticated processing of these data possible. The name **chemometrics** has been coined to describe mathematical and statistical methods designed to extract useful chemical information from chemical data. Optimization is one field covered by the umbrella of chemometrics; another is the field of pattern recognition, to which we now turn.

One consequence of automation is that many parameters can be determined simultaneously for the same sample, for example in clinical chemistry, chromatography and atomic emission analysis. One use of such data in analytical chemistry is significance testing, e.g. determining whether an oil spill comes from a particular source, by using the intensities and/or frequencies of the many maxima obtained in the infrared or fluorescence spectra of such samples. Another use is classification, e.g. dividing the stationary phases used in gas–liquid chromatography into groups with similar properties, by using the retention properties of a variety of solutes with different chemical properties. In both cases pattern recognition methods use the available data simultaneously rather than sequentially. The set of measurements which is used to characterize the sample is called a **pattern**. When only two parameters are measured for each sample the pattern can be represented graphically by a point, as shown in Fig. 6.12 where the coordinates of the point are the values taken by the two paramters. The point can also be defined by a vector, drawn to it from the origin and known as a **pattern** or **data vector**; the coordinate system is known as the **pattern space**. The basis of all pattern-recognition methods is that pattern vectors for similar samples lie close together in the pattern space, forming clusters. In two dimensions this clustering can easily be detected by the human eye. However, when more than two parameters are measured, each sample will be represented by a point in $n$-dimensional space and mathematical methods are needed to detect clustering. The distance, $d$, between two points in $n$-dimensional space with coordinates $(x_1, x_2, \ldots x_n)$ and $(y_1, y_2, \ldots y_n)$ is usually taken as the Euclidian distance, calculated by using the expression:

$$d^2 = \sum_{i=1}^{n} (x_i - y_i)^2$$

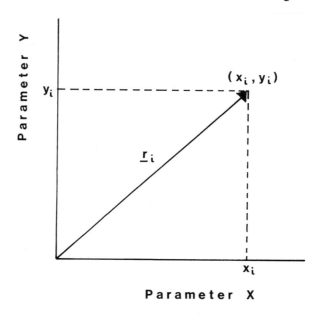

Fig. 6.12 – A pattern represented in pattern space; $x_i$ and $y_i$ are the values of the parameters X and Y; $r_i$ is the pattern vector.

There are essentially two different approaches to pattern recognition, depending on whether it is being used for significance testing or classification. In the former case the classes into which a sample may fall are known, and the purpose is to enable a sample of unknown class to be classified by its pattern. In the case of the oil-spill example, a sample may have come from one of several different sources: each source forms a class and the purpose of pattern recognition is to identify the class and hence the source of the sample. In the second case the purpose is to see whether the patterns obtained fall into natural groups; in this case there is no prior knowledge of the classes to be expected. These two different approaches to pattern recognition are sometimes called **supervised** and **unsupervised learning**, respectively. They use different methods, to which we now turn.

## 6.14  SUPERVISED LEARNING METHODS

The simplest situation is that in which a sample may belong to one of two classes: this is known as *binary classification*. It is illustrated in Fig. 6.13 for the two-parameter case. In order to find a criterion or **classifier** for separating the two classes, a data set of patterns for samples with known class membership is used. This data set is divided (by using random numbers) into two parts

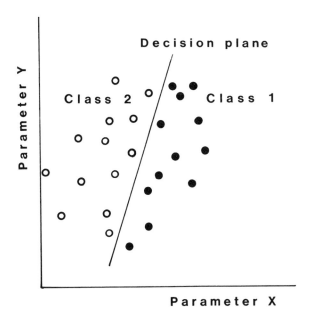

Fig. 6.13 – Two classes separated by a decision plane.

called the **training set** and the **prediction set**. The training set is used to find the position of a plane, the **decision plane**, which separates the two classes. A simple way of doing this is to use the plane of symmetry between the centres of gravity of the two classes. This means that an unknown pattern vector is assigned to the class which has its centre of gravity closest to the vector. This method is satisfactory if the classes form compact clusters, but if they do not, more effective separation can be obtained by using more sophisticated methods. One of these is the **learning-machine method**. This is an iterative procedure using negative feed-back. Starting with the plane of symmetry between the two centres of gravity as the decision plane, the patterns of the training set are tested one by one. If an incorrect classification is obtained, the decision plane is adjusted to rectify this. The method is called the learning-machine method because it 'learns' from the mistakes in classification what adjustment is required in the position of the decision plane. Provided that two classes *can* be separated by a plane, this procedure eventually leads to the position of such a plane. If the classes are *not* separable by a plane, then the least-squares method is preferable. This method minimizes a sum of squares which estimates the errors in classification. Determining the position of the decision plane can also be treated as an optimization problem, solvable by the simplex method. The decision plane can be defined by a vector orthogonal to it and the components of this vector are the factors to be varied in level. The response which is maximized is the proportion of the patterns from the training set which have been correctly classified. The effectiveness of the decision plane in separating the two classes can be tested by using the remainder of the data set, i.e. the prediction set: each sample is classified by its pattern, and the percentage correctly classified gives an objective estimate of the efficiency of the classifier.

In practice, before the pattern-recognition procedure is applied the data will be given a preliminary examination, for two closely related reasons. It may be possible to exclude altogether one or more parameters which clearly make little or no contribution to the efficiency of the classifier: indeed, if too many parameters are used, a random and chemically meaningless separation into classes may result. Additionally, the different parameters (which may be measured in different numerical values) can be scaled, and a preliminary attempt may be made to weight the parameters to optimize the classification. After the classifier has been tested, these weighting factors may be further revised.

When there are more than two classes to be separated, a series of binary classifiers may be used. An alternative method which is conceptually very simple is the **_K_ nearest-neighbour (KNN) method**. This method also has the advantage that it can be used when the classes cannot be separated by a plane, as illustrated in Fig. 6.14. In the simplest form of separation each sample is classified as in the

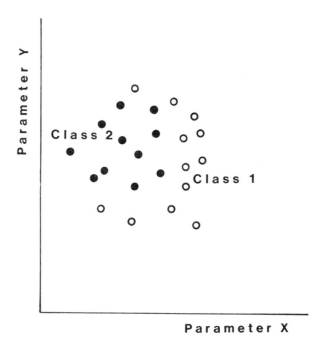

Fig. 6.14 – Two classes which cannot be separated by a plane

same class as its nearest neighbour. A more spohisticated method uses the $K$ nearest neighbours (where $K$ is a small integer) and class membership is decided by a voting scheme, possibly with different weightings given to the neighbours, depending on their relative distances from the sample. With this method no division of the data set into training and prediction sets is necessary — all the data can be used to estimate the efficiency of the method.

The approaches to pattern recognition so far described have been non-parametric, i.e. no model is proposed for the distribution of samples within a class. In contrast, the commercially available SIMCA computer program (due to Wold) uses the data set to form a mathematical model of each class. It is then possible to calculate for each class the probability that an unknown sample belongs to that class.

Some of the fields in which supervised pattern-recognition methods have been applied are diagnostic tests in clinical chemistry, forensic science, the identification of sources of pollutants, food chemistry and the prediction of molecular structure from spectra.

## 6.15 UNSUPERVISED LEARNING METHODS

The reader will remember that these are methods which are used to decide whether a set of patterns divides naturally into groups. There are a large number of methods of searching for clusters in pattern space. One approach finds the pair of points which are closest together and replaces them by a new point half way between. This procedure is repeated, and if continued indefinitely will group all the points together as shown by the steps in Fig. 6.15. In these diagrams the pairs of points which have been paired together and replaced by a single point are joined by a line. Thus at any stage groups of points which are classed together are joined. The stage at which the grouping is stopped, i.e. the final classification, should be decided on a chemical basis. An opposite approach treats all the patterns as one group initially and then subdivides it.

As mentioned above, unsupervised learning methods have been used to classify the many stationary phases used in gas-liquid chromatography. A small preferred set of phases with different characteristics can then be selected by taking one phase from each class. Another application is the classification of antibiotics in terms of their activity against various types of bacteria, in order to elucidate the relationship between biological activity and molecular structure.

It will be seen that pattern recognition can be used to establish relationships which might otherwise be concealed in a mass of data. It is a method which has only recently been applied in analytical chemistry: its potential as a means of data analysis remains to be fully evaluated. A number of computer packages are available, besides SIMCA. One such is ARTHUR (developed by Kowalski) which contains programs for many methods, including those described here. SIMCA and ARTHUR are described in more detail in a reference given in the bibliography to this chapter.

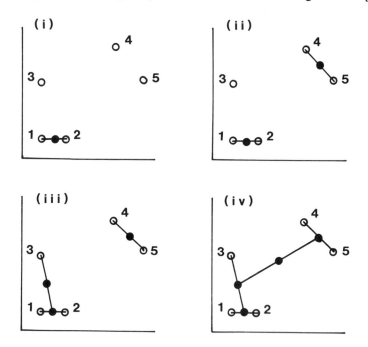

Fig. 6.15 — Stages in the clustering of patterns: o represents an original point, • represents the single point replacing a pair of points.

## 6.16 SAMPLING

The final section of this book discusses the point at which the analyst starts work — the taking of samples. Although this topic has been left until last it is of fundamental importance. Unless the sampling stage is given careful consideration the statistical methods previously described are of little use.

An analyst has to deal with samples, since in most cases it is impracticable or impossible to analyse all of the object under consideration. For example it is not practicable to analyse a tankerful of milk for the fat content and impossible to analyse all the water in a river for a particular pollutant. Furthermore many analytical methods destroy the object which is being analysed and so cannot be applied to the whole of an object of value.

To illustrate some aspects of sampling let us consider the situation in which we have a large batch of tablets and wish to obtain an estimate for the mean weight of a tablet. Instead of weighing all the tablets, we take a few, say ten, and weigh each of them. In this example the batch of tablets forms the population and the ten tablets weighed form a sample from this population (see Section 2.2). If the sample is to be used to deduce the properties of the population, it must be what is known statistically as a **random sample.** This is a sample taken in such a way that all the members of the population have an equal chance of being included. Only then will equations such as (2.9), which gives the confidence limits of the mean, be valid. It must be appreciated that the term 'random' has, in the statistical sense, a meaning different from 'haphazard'. Although in practice an analyst might spread the tablets on his desk and attempt to pick a sample of ten in a haphazard fashion, such a method could conceal an unconscious bias. The best way to obtain a random sample is by the use of a random number table. The method is similar to that described in Section 6.2 for randomizing the order of experiments. Each member of the population is allocated a number (such that all the numbers have an equal number of digits) and then numbers are taken from a random number table, starting at an arbitrary point, in order to select the sample. An alternative (and simpler) method which is sometimes used is to select the population members at regular intervals, for example to take every one hundredth tablet off a production line. This method is not entirely satisfactory, since there might be a coinciding periodicity in the weight of the tablets.

A random sample of $n$ tablets having been obtained the weight of each tablet can be measured and the mean, $\bar{x}$, and the variance, $s^2$, of these weights calculated. Since the confidence limits of the mean are given by Eq. (2.9):

$$\mu = \bar{x} \pm t(s/\sqrt{n})$$

we require $s$ to be as small as possible. There are two contributions to $s^2$: the **sampling variance,** $\sigma_1^2$, due to differences in weight between the tablets, and the **measurement variance,** $\sigma_0^2$, due to random errors in the measurement of weight. Section 3.11 explained how these two contributions can be separated and estimated by using ANOVA. The total variance of the measured weights, which is estimated by $s^2$, is the sum of the sampling and measurement variance, i.e. $\sigma_0^2 + \sigma_1^2$ (see Sections 1.6 and 2.7). If the weight of each tablet is given by the mean of $m$ replicate weighings, $s^2$ becomes $\sigma_0^2/m + \sigma_1^2$. The term due to the measurement variance can be reduced either by using a more sensitive balance or by increasing $m$, the number of replicate weighings of each tablet. However, there is no point in striving to make it less than say a tenth of the sampling variance, since any further reduction will not greatly improve the total variance (which is the sum of the two variances). Rather it is preferable to take a larger sample, since the confidence interval decreases with increasing $n$. If a preliminary sample has been used to estimate $s$, then the sample size required to achieve a given size of confidence interval can be calculated approximately.

In the example above the population is made up of obvious discrete members which are nominally the same, i.e. the tablets. Sampling from materials for which this is not true, such as rocks, powders, gases and liquids, is called **bulk sampling.** The lack of identifiable units leads to some confusion in nomenclature — should 'sample' take its colloquial meaning of 'a small part of the whole' so that *one* sample corresponds to *one* tablet in the previous example, or should it take the statistical meaning of several small parts taken from different parts of the bulk so that *one* sample corresponds to *several* tablets? In order to avoid this confusion in the following discussion, a small part taken from the bulk will be called a **sample increment** and a group of such sample increments will be called a **gross sample**.

If a bulk material were perfectly homogeneous then only one sample increment would be needed to determine the properties of the bulk. In practice, bulk materials are non-homogeneous for a variety of reasons. Materials such as ores and sediments consist of macroscopic particles with different compositions and these may not be uniformly distributed in the bulk, and fluids may be non-homogeneous on a molecular scale due to concentration gradients. Such inhomogeneity is only detected by taking sample increments from different parts of the bulk. In order to obtain a random gross sample, sample increments should be taken at points in the bulk which are chosen at random. This can be done by considering the bulk as a collection of cells and using random numbers to select the cells to be sampled, as described above.

The sampling variance depends on the size of the sample increment relative to the scale of the inhomogeneities and decreases with increasing sample increment size. In some experiments it may be necessary to set an upper limit on the sampling variance so that changes in the mean can be detected. Preliminary measurements can be made to decide the minimum size of sample increment required to give an acceptable level of sampling variance.

A possible sampling strategy with bulk material is to take $n$ sample increments and blend them before making $m$ replicate measurements. The variance of the mean of these replicate measurements is $\sigma_0^2/m + \sigma_1^2/n$ (where $\sigma_0^2$ is the measurement variance and $\sigma_1^2$ the sampling variance, as already defined above). This total variance should be compared with that for the mean when each sample increment is analysed $m$ times and the increment means are averaged, which gives a variance $(\sigma_0^2/m + \sigma_1^2)/n = \sigma_0^2/mn + \sigma_1^2/n$. Obviously the latter variance is the smaller, resulting in greater precision of the mean, but more measurements ($mn$ against $m$) are required. Knowledge of the values of $\sigma_0^2$ and $\sigma_1^2$ from previous experience, and the costs of sampling and analysis, can be used to calculate the relative costs of different sampling strategies. In general the most economical scheme to give the required degree of precision will be used.

## BIBLIOGRAPHY

G. E. P. Box, W. G. Hunter and J. S. Hunter, *Statistics for Experimentalists,* Wiley, New York, 1978. Gives a detailed treatment of complete and fractional factorial designs and the method of steepest ascent.
R. Caulcutt and R. Boddy, *Statistics for Analytical Chemists,* Chapman & Hall, London, 1983. Gives a very detailed treatment of the use of ANOVA in interlaboratory collaborative experiments for the standardizing of test methods.

B. E. Cooper, *Statistics for Experimentalists,* Pergamon Press, Oxford, 1975. Give further details of the mathematics of ANOVA.

D. R. Cox, *Planning of Experiments,* Wiley, New York, 1958. A non-mathematical treatment intended for the experimental worker. Gives further detail about Latin squares.

O. L. Davies and P. L. Goldsmith, *Statistical Methods in Research and Production,* Longmans, London, 1972. Gives more detail on the separation of variance components, confidence limits for estimates of variance, and efficiency of different experimental designs.

G. Kateman and F. W. Pijpers, *Quality Control in Analytical Chemistry,* Wiley, New York, 1981. Gives a very detailed treatment of sampling.

B. R. Kowalski, (ed.), *Chemometrics: Theory and Application,* American Chemical Society, Washington, 1977. Contains articles on SIMCA and ARTHUR.

D. L. Massart, A. Dijkstra and L. Kaufman, *Evaluation and Optimisation of Laboratory Methods and Analytical Procedures,* Elsevier, Amsterdam, 1978. A valuable source of information on all topics in this chapter. Includes many examples drawn from the literature and gives a large number of further references.

K. Varmuza, *Pattern Recognition in Chemistry,* Springer-Verlag, Berlin, 1980. Gives a detailed treatment of the methods and applications of pattern recognition.

L. A. Yarbro and S. N. Deming, *Anal. Chim. Acta,* 1974, **73,** 391. Discusses the choice of the initial simplex.

## EXAMPLES

1. In an interlaboratory collaborative experiment on the determination of arsenic in coal, samples of coal from three different regions were sent to each of three laboratories. Each laboratory performed a duplicate analysis on each sample, with the results shown below (measurements in $\mu$g/g).

|        |          | Laboratory |          |
|--------|----------|----------|----------|
| Sample | 1        | 2        | 3        |
| A      | 5.1, 5.1 | 5.3, 5.4 | 5.3, 5.1 |
| B      | 5.8, 5.4 | 5.4, 5.9 | 5.2, 5.5 |
| C      | 6.5, 6.1 | 6.6, 6.7 | 6.5, 6.4 |

Verify that there is no significant laboratory–sample interaction and test for significant differences between the laboratories.

2. If the response at vertex 7 in the example on simplex optimization (p. 163) is found to be 12, which vertex should be rejected in forming the new simplex and what are the coordinates of the new vertex?

3. Four standard solutions were prepared, each calculated to contain 16.00% (by weight) of chloride. Three titration methods, each with a different technique of end-point determination, were used to analyse each standard solution. The order of the experiments was randomized. The results for the chloride found (% w/w) are shown below

|          |       | Method |       |
|----------|-------|-------|-------|
| Solution | A     | B     | C     |
| 1        | 16.03 | 16.13 | 16.09 |
| 2        | 16.05 | 16.13 | 16.15 |
| 3        | 16.02 | 15.94 | 16.12 |
| 4        | 16.12 | 15.97 | 16.10 |

Test whether there are significant differences between (a) the concentration of chloride in the different solutions and (b) the results obtained by the different methods.

4. Samples of four different boiler feeds were analysed for nitrovin content by two different methods, each in duplicate. The results are shown in the table below (concentrations in mg/kg). Test whether the results obtained by the two different methods differ significantly.

| Feed | HPLC Method | Spectrophotometric method |
|------|-------------|---------------------------|
| 1 | 9.3, 10.0 | 9.0, 10.2 |
| 2 | 9.3, 12.3 | 9.0, 11.8 |
| 3 | 12.8, 12.7 | 12.9, 12.0 |
| 4 | 11.9, 11.8 | 11.6, 12.1 |

(Adapted from M. J. Gliddon, C. Cordon and G. M. Parnham, *Analyst,* 1983, **108,** 116).

5. The effective temperature, calculated from the absorbance ratio of two spectral lines, for tin vapour was compared with the actual temperature under different experimental conditions in a heated quartz tube. The results are shown in the table below (values in °C). Calculate the main and interaction effects.

Effective temperature (°C)

|  |  | Purge gas Ar | | Purge gas Ar + 1% $O_2$ | |
|------|------|------|------|------|------|
| Purge time (sec) |  | 20 | 60 | 20 | 60 |
| Actual temperature (°C) | 900 | 843 | 826 | 852 | 855 |
|  | 1000 | 908 | 873 | 908 | 908 |

(Adapted from B. Welz and M. Melcher, *Analyst,* 1983, **108,** 213).

6. Mercury is lost from solutions stored in polypropylene flasks by combination with adsorbed tin. The absorbance of a standard aqueous solution of mercury stored in such flasks was measured for two levels of the following factors:

| Factor | Low | High |
|--------|-----|------|
| A — agitation of flask | Present | Absent |
| C — cleaning of flask | Once | Twice |
| T — standing time | 1 hr | 18 hr |

The following results were obtained. Calculate the main and interaction effects.

| Combination of factor levels | Absorbance |
|------------------------------|------------|
| 1 | 0.084 |
| a | 0.099 |
| c | 0.082 |
| t | 0.049 |
| ac | 0.097 |
| ta | 0.076 |
| tc | 0.051 |
| atc | 0.080 |

(Adapted from A. Kuldvere, *Analyst,* 1982, **107,** 179).

7. The percentage recovery of lead in spiked samples in the presence of 'low' and 'high' concentrations of nitric acid and lanthanum was investigated by electrothermal atomic-absorption spectroscopy; the means of duplicate analyses are given below. Calculate the main and interaction effects. If the residual mean square is 4 (with 4 degrees of freedom) test whether the interaction effect is significant.

| | Lanthanum concentration (% w/v) | |
|------|------|------|
| Nitric acid concentration (% v/v) | 0 | 0.10 |
| 0.1 | 64 | 90 |
| 1.0 | 77 | 96 |

(Adapted from M. P. Bertenshaw, D. Gelsthorpe and K. C. Wheatstone, *Analyst,* 1982, **107,** 163).

8. Two sampling schemes are proposed for a situation in which it is known, from past experience, that the sampling variance is 10 and the measurement variance 4 (arbitrary units).

Scheme 1:  take 5 samples, blend them and perform a duplicate analysis.
Scheme 2:  take 3 samples and perform a duplicate analysis on each.

Show that the variance of the mean is the same for both schemes.
What ratio of the cost of sampling to the cost of analysis must be exceeded for the second scheme to be the more economical?

# Solutions to exercises

## CHAPTER 1

1. Laboratory A has achieved a mean result of 41.9 g/l, (very close to the correct value) and a small spread of results — all the values lie between 41.1 and 42.5 g/l. These results are thus accurate and precise: both random and systematic errors are small. Laboratory B has achieved the same mean value, but the spread of the results is much larger (range 39.8–43.9 g/l.), so although there are apparently no systematic errors (i.e. the mean result is accurate) there are large random errors (i.e. the data are very imprecise). Similar considerations show that laboratory C has produced precise but inaccurate results, (mean 43.2 g/l.), and that the results from laboratory D are both imprecise and inaccurate. Laboratory E has produced a set of results that look both precise and accurate, with the exception of the final value. In practice, this value might be tested as an outlier (see Chapter 3): if the tests showed that it could, with a reasonable level of confidence, be rejected, the remaining results would be quite similar to those of laboratory A.

2. The second set of six results obtained by Laboratory A have the same mean result as the first set, confirming that this laboratory produces results with no significant systematic error. In the second set the spread is greater (range 40.8–43.3 g/l.), illustrating the difference between within-day or within-batch precision (repeatability) and between-day or between-batch precision (reproducibility).

3. Monoclonal antibody populations are homogeneous, hence the number of binding sites per molecule must be a whole number — obviously two in this case. The results are thus precise, but show clear evidence of a systematic error which produces low values.

4. (i) The lactate concentration in human blood varies quite widely from one healthy subject to another (ca. 5–20 mg/100 ml), and also varies to a lesser extent within one subject from time to time. If the degree of the latter type of variation is studied, accuracy will not be very important but precise measurements will be necessary — the experimental errors must be negligible compared with the variations for the individual. If a single measurement is made, to check

whether a subject does or does not fall within the the 'normal range' for blood lactate, less precision will be required, but a large systematic error might lead to a wrong diagnosis.

(ii) The uranium content of ores is studied with a view to economically viable extraction of the element. High precision will not therefore be necessary, but substantial systematic errors (positive or negative) might lead to economically disastrous decisions!

(iii) In this analysis speed is clearly essential, and neither precision nor accuracy is of great importance initially. As the poisoned patient recovers, the plasma level of the drug may be repeatedly monitored to confirm that it is decreasing: since a trend is being studied, precision is more important than accuracy.

(iv) Here again, the main aim is to detect changes in the analytical result; as these changes may be quite small, good precision will be necessary to detect any trend, but accuracy will be less essential.

5. (i) In this experiment the most likely source of error is that the sample taken is not representative of the ore as a whole, and may thus give a wholly misleading value for the bulk Fe content (sampling is discussed in Chapter 6). A systematic error will also arise if the reduction of Fe(III) to Fe(II) is incomplete, or if there is a large indicator error. Systematic errors other than the sampling error can be checked with the aid of a standard ore sample — such samples are commercially available and are supplied with a certified Fe value. An additional problem, not necessarily solved by using the standard sample, is the possibility that other elements in low oxidation states will be titrated by the ceric sulphate, giving a falsely high result for iron. Random errors in volumetric analysis are discussed in the earlier sections of this Chapter.

(ii) Apart from the systematic errors discussed in (i), incomplete chelate formation and/or extraction will present the major problems in this case. Again, such errors will be detectable with the aid of an ore sample of known Fe content. The colorimetric analysis will probably use a series of Fe standards and a graphical procedure to calculate the result and the random errors: such methods are further discussed in Chapter 4.

(iii) Random errors in gravimetric analysis should be very small and the systematic errors directly associated with the weighing process can be minimized by careful technique (see Section 1.4). The most likely sources of errors in this determination are chemical e.g. the problem of co-precipitation of other ions, and are discussed at length in the established textbooks of classical analysis.

## CHAPTER 2

1. Mean = 0.077 $\mu g/g$. Standard deviation = 0.007 $\mu g/g$. RSD = 9%.

2. Mean = 5.163. Standard deviation = 0.027. Number of degrees of freedom = 6 (i) 5.163 ± 0.025 (ii) 5.163 ± 0.038

3. $20.9 + 0.3$ g/l.

4. $10.1 \pm 0.17$ ng/ml. Since the sample is large, take $t = 1.96$. Sample size is calculated from $0.1 = 1.96 \times 0.6/\sqrt{n}$, giving $n = 138$.

5. Mean $= 10.45\%$. Standard deviation $= 0.105\%$. 95% confidence limits $= 10.45 \pm 0.11\%$. 99% confidence limits $= 10.45 \pm 0.17\%$. (i) No (ii) No.

6. Mean $= 1.74$, standard deviation $= 1.28$ g/l. The logarithms of the given values are: 0.260, 0.521, 0.029, 0.104, $-0.309$, 0.579, $-0.824$, 0.297. The mean of these values is 0.082 and their standard devaition is 0.462. The 95% confidence interval for the logarithmic values is $0.082 \pm 0.386$, i.e. from $-0.304$ to 0.468. Taking antilogarithms gives the confidence interval of the mean as 0.50-2.94.

7. Mean $= 0.2943$. Standard devition $= 0.00468$. 99% confidence limits $= 0.294 \pm 0.005$.

8. Mean $= 10.186$ ml. Standard deviation $= 0.188$ ml. 95% confidence limits $= 10.186 \pm 0.233$, i.e. 9.95-10.42 ml. Since the confidence interval includes 10.00 ml there is no evidence of systematic error.

9. Standard deviation $= 0.14$ mg. Weight of reagent required is 0.5(4.9) g, giving RSD $= 0.028\%$ (0.0029%). RSD for volume of solvent $= 0.02\%$. RSD of molarity of solution $= 0.034\%$ (0.020%). Values for reagent with formula weight 392 are given in brackets.

10. Solubility $= \sqrt{1.3 \times 10^{-10}} = 1.1_4 \times 10^{-5} M$. RSD of solubility product $= 7.7\%$ giving RSD of solubility $= 3.85\%$ and standard deviation $= 0.04_4 \times 10^{-5} M$.

## CHAPTER 3

1. Points lie approximately on straight line.

2. $Q = 0.565$. Critical value $= 0.570$.

3. Tomato: mean $= 772.6$, standard deviation $= 13.6$ $\mu$g/g. Cucumber: mean $= 780.9$, standard deviation $= 10.4$ $\mu$g/g.
(i) $F_{6,6} = 1.7$. Critical value $= 5.820$ ($P = 0.05$). Variances do not differ significantly.
(ii) Pooled estimate of standard deviation $= 12.1$ $\mu$g/g, $t = 1.28$. Critical value $= 2.18$ ($P = 0.05$). Means do not differ significantly.

4.

| Source of variation | Sum of squares | Degrees of freedom | Mean square |
|---|---|---|---|
| Between samples | 6365.7116 | 3 | 2121.9 |
| Within samples | 162.0284 | 20 | 8.1 |
| | 6527.74 | 23 | |

$F_{3,20} = 262$. Critical value $= 3.098$ ($P = 0.05$). Sample means: 34.1, 45.4, 72.2, 70.4. Least significant difference $= 3.43$. All pairs of levels, except the deepest pair, differ significantly from each other.

5. Use the chi-squared test with each expected frequency = 5. $\chi^2$ = 16.8. Critical value = 16.92 ($P = 0.05$). No evidence that some digits are preferred.

6. $t$ = 1.54, 1.60, 1.18, 160. Critical value = 2.36 ($P = 0.05$). None of the means differs significantly from certified value.

7. Means = 55.9 and $50.2_5$ mg/kg, Standard deviations = 0.834 and 8.83 mg/kg. (a) $F_{7,7}$ = 112. Critical value = 3.787 ($P = 0.05$). Variability significantly greater for longer boiling time.
(b) $t$ = 1.8, 7 degrees of freedom. Critical value = 2.36 ($P = 0.05$). Means do not differ significantly.

8. (a) Expected frequencies = 15.25, 45.75. $\chi^2$ = 5.95 (Yates' correction applied). Critical value = 3.84 ($P = 0.05$). First worker differs significantly from other three.
(b) Expected frequencies = 12.33, 12.33, 12.33. $\chi^2$ = 2.81. Critical value = 5.99 ($P = 0.05$). Last three workers do not differ significantly from each other.

9. First test whether variances differ significantly: $F_{2,2}$ = 1.22 − not significant. $t$ = 0.56. Critical value = 2.78 ($P = 0.05$). Mean results do not differ significantly.

10. Paired $t$-test. Differences are 0.051, 0.030, −0.030 mg/ml. Mean = 0.017 mg/ml, standard deviation = 0.042 mg/ml. $t = 0.7$. Critical value = 4.3 ($P = 0.05$). Results do not differ significantly.

11. Men: mean = 40.0 g/l, standard deviation = 2.78 g/l. Women: mean = $43.2_5$ g/l, standard deviation = 3.06 g/l. Variances do not differ significantly. $t$ = 2.2. Critical value = 2.14 ($P = 0.05$). Means differ significantly.

12. Paired $t$-test. $t$ = 3.4. Critical value = 2.57 ($P = 0.05$). Means differ significantly.

13.

| Source of variation | Sum of squares | Degrees of freedom | Mean square |
|---|---|---|---|
| Between days | 332.92 | 3 | 111 |
| Within days | 26.00 | 8 | 3.25 |
| Total | 358.92 | 11 | |

$F_{3,8}$ = 34. Critical value = 4.066 ($P = 0.05$). Mean concentrations differ significantly. Estimate of sampling variance = 35.9.

# CHAPTER 4

1. Application of Eq. (4.2) to the data gives a product-moment correlation coefficient, $r$, of −0.8569. This value suggests a considerable negative correlation between the mercury level and the distance from the polarograph. Its significance can be tested by using Eq. (4.3), which yields a $t$-value of 3.33. Comparison of this value with the tabulated value of $t$ at the 95% confidence level and $n - 2$

(= 4) degrees of freedom ($t = 2.78$) shows that the relationship is indeed significant at this confidence level. Two cautions, however: (1) this method tests only linear correlation, whereas in the laboratory we might expect that the mercury level would decrease with increasing distance from the polarograph in an inverse square manner; (2) correlation is *not* the same as causation. The calculation shows that there is a relationship between distance from the polarograph and mercury concentration – it does *not* show that the polarograph is the source of the mercury contamination. The contamination might arise (for example) from a sink or another piece of equipment close to the polarograph.

2. In this case Eq. (4.2) shows that $r = 0.99982$. This is apparently strong evidence for linear correlation, *but* inspection of the data shows clearly that a curve is more suitable in this case. Each $y$ (absorbance) value is greater than its predecessor, but by a slowly decreasing increment as $x$ increases. The curvature can also be demonstrated by studies of the residuals (cf. exercise 9). In this case, the errors caused by treating the data as a linear calibration plot would be very small.

3. Application of Eqs. (4.4) and (4.5) gives $b = 0.0252$, and $a = 0.0021$ respectively. Other important results include: $\bar{x} = 15$; $\bar{y} = 0.380$; $\Sigma x_i^2 = 2275$; and $\sum_i (x_i - \bar{x})^2 = 700$. The individual $y$ residuals ($y_i - \hat{y}$) are found to be +0.0009, −0.0009, −0.0028, +0.0104, −0.0074, −0.0062, +0.0060. These residuals sum (as expected) to zero, and the sum of their squares is 0.000247. Application of Eq. (4.6) shows that $s_{y/x} = 0.00703$. Then Eq. (4.7) gives $s_b = 0.000266$, and the use of $t$ (5 degrees of freedom; 95% confidence level) = 2.57 yields for $b$ confidence limits of $0.0252 \pm 0.0007$. Similarly, Eq. (4.8) yields $s_a = 0.00479$, with confidence limits for $a$ of $0.0021 \pm 0.0123$.

4. (a) It is easy to show, using the $a$ and $b$ values calculated in exercise 3, that a $y$ value of 0.456 corresponds to an $x$ value of 18.04 ng/ml, and that $(y_0 - \bar{y})^2 = 0.00578$. The application of Eq. (4.9) then gives $s_{x_0} = 0.300$, and the corresponding confidence limits are given by $18.04 \pm (2.57 \times 0.300) = 18.04 \pm 0.77$ ng/ml.

(b) In this case it is clearly necessary to test to see whether the absorbance value 0.347 is an outlier. Form Eq. (3.8) the calculated value of $Q$ is (0.347−0.314)/(0.347 − 0.308) = 0.846. At the 95% confidence level, this just exceeds the tabulated $Q$ value of 0.829, so the value 0.347 *can* be rejected. This leaves three results, with a mean absorbance of 0.311, corresponding to a concentration of 12.28 ng/ml. In this case the value of $s_{x_0}$ is given by Eq. (4.10) ($m = 3$; $(y_0 - \bar{y})^2 = 0.00476$) as 0.195, the confidence limits thus being $12.28 \pm (2.57 \times 0.195) = 12.28 \pm 0.50$ ng/ml.

5. Armed with the results of exercise 3 above, this calculation is very simple. The limit of detection is defined as the concentration yielding a signal exceeding the background absorbance signal by three standard deviations, the background

being given by $a$ (0.0021) and the standard deviation by $s_{y/x}$ (0.00681). The absorbance signal at the limit of detection is thus $0.00211 + (3 \times 0.00681) = 0.0225$, and the limit of detection, calculated from the slope and intercept of the regression line, is 0.81 ng/ml.

6. The usual linear regression calculation gives $a = 0.2569$ and $b = 0.005349$. The ratio of these two numbers gives the concentration of the test sample — 48.0 ng/ml. Also, $\bar{x} = 35$, $\bar{y} = 0.4441$, $\Sigma(x_i - \bar{x})^2 = 4200$, and $\Sigma(y_i - \hat{y})^2 = 0.00001364$. Thus $s_{y/x} = 0.003693$, and $s_{x_e}$, given by Eq. (4.12) is 0.9179. The confidence limits for the concentration are thus $48.03 \pm (2.45 \times 0.9179) = 48.0 \pm 2.2$ ng/ml.

7. The data provided allow a weighted regression calculation, the mean $y$ values and standard deviations being:

| $x$: | 0 | 10 | 20 | 30 | 40 | 50 |
|------|------|------|------|------|------|------|
| $y$: | 4.0 | 21.2 | 44.6 | 61.8 | 78.0 | 105.2 |
| $s$: | $0.7_1$ | $0.8_4$ | $0.8_9$ | $1.6_4$ | $2.2_4$ | $3.0_3$ |

The first step is to calculate the unweighted regression line, the slope and intercept being 1.982 and 2.924 respectively. The concentrations corresponding to fluorescence intensities of 15 and 90 units are thus 6.09 and 43.9 ng/ml respectively. In the usual way (cf. exercise 3 above), $s_{y/x}$ is calculated to be 2.990. Since $\Sigma(x_i - \bar{x})^2$ is readily calculated as 1750 and $\bar{y}$ as 52.47, the $s_{x_0}$ values for the two concentrations can be shown [Eq. (4.9)] to be 1.767 in each case. By use of a $t$ value of 2.78, the 95% confidence limits for these values are thus given by $6.1 \pm 4.9$ and $43.9 \pm 4.9$ ng/ml respectively.

To calculate the weighted regression line, it is first necessary to calculate the weight for each point by using Eq. (4.13). In order of increasing $x_i$ value, these weights are found to be 2.227, 1.591, 1.418, 0.417, 0.224, and 0.122 – note that the sum of these numbers is 5.999, i.e. 6 if we allow for rounding errors. This is an important check to make. By use of these weights and Eqs. (4.14) and (4.15) the slope and intercept of the weighted regression line are found to be 1.964 and 3.483 respecitvely. From this line, the concentrations corresponding to fluorescence intensities of 15 and 90 units are 5.87 and 44.1 ng/ml respectively. Note that these values, and those for the slope and intercept, are all similar to those calculated for the unweighted line, despite the considerable scatter of the experimental points. The crucial difference between the use of the weighted and unweighted calculation methods only becomes apparent when Eq. (4.16) is used to estimate the standard deviations of the estimated concentrations. To use this equation, it is first necessary to estimate $w_0$ values for the two $y_0$ values under study. Inspection of the data suggests that weights of 1.8 and 0.18 respectively should be appropriate for the $y_0$ values of 15 and 90. Use of these weights gives $s_{x_{0w}}$ values of 0.906 and 2.716 respectively. Multiplication by the $t$ value 2.78

gives the confidence limits for the two concentrations as $5.9 \pm 2.5$ and $44.0 \pm 7.6$ ng/ml respectively.

8. Here, the "new" method is clearly the ISE one, plotted therefore on the $y$-axis: the gravimetric technique is the reference or standard procedure and is thus plotted on the $x$-axis. The regression line calculated in this way has a slope of 0.963, and an intercept of 4.48, the correlation coefficient being 0.970. This last value is clearly highly significant $[t = 11.29 - $ Eq. (4.3)], but the confidence limits of the slope and intercept must also be calculated. The usual calculations [Eqs. (4.6)–(4.8)] in conjunction with a tabulated $t$ value of 2.31 (8 degrees of freedom; 95% confidence) show that the confidence limits for the slope and the intercept are respectively $0.96 \pm 0.20$ and $4.5 \pm 20.1$. Since these 95% confidence intervals clearly include 1 and 0 respectively, we must conclude that this regression line indicates good agreement between the two analytical methods. Nonetheless, there is a residual feeling of unease about the data. Calculation shows that 7 of the $y$-residuals are negative (including 6 in a single sequence) and only 3 positive. There is also a suggestion that the agreement between the two methods is very close for most of the samples, but very poor for a few of them; this could be caused by speciation problems. A good analyst might conclude that more data are needed (more are actually supplied in the paper cited).

9. If these data are plotted graphically (remember that this should *always* be done), there is some indication that the calibration graph is linear up to $0.7 - 0.8$ absorbance units, and curved thereafter. Unweighted linear regression calculations using all six points yield a correlation coefficient of 0.9936, and $y$-residuals (in order of increasing $x$) of $-0.07, -0.02, +0.02, +0.06, +0.07,$ and $-0.07$ (rounded to 2 places of decimals). The trend in these values is obvious and confirms that we are dealing with a curve. The sum of squares of the residuals is 0.0191. Confirmation that the last point should be omitted from the linear range is obtained by repeating the regression calculation with only the first five points. The correlation coefficient then increases to 0.9972 and the $y$-residuals are $-0.04, 0, +0.02, +0.04,$ and $-0.02$ respectively (sum of squares 0.0040). The $y$-residual for the sixth point with respect to this second straight line is very large by comparison: $-0.22$.

   This process can then be repeated for the fifth point, i.e. the straight line fitted to only the first four points is calculated. Again there is an improvement in the correlation coefficient (to 0.9980), a reduction in the values of the $y$-residuals $(-0.02, +0.01, +0.01,$ and $-0.01)$, and a large residual for the fifth point under test $(-0.14)$. It can thus be concluded that the fifth point can also be excluded from the linear range of the method. Similar tests applied to the fourth point indicate by contrast that it probably should be part of the linear range, and an analyst could use linear regression methods in the absorbance range 0–0.6 with confidence.

10. This example also emphasises the importance of drawing a calibration curve rather than simply performing a linear regression calculation. The latter calculation yields a correlation coefficient of 0.9952, yet inspection of the plot shows that it is obviously curved. Linear regression shows the $y$-residuals have signs (in order of increasing $x$) of $-\,-\,-\,+\,+\,+\,+\,+\,+\,-\,-$: the sum of the squares of the residuals is 9.50. The order of the signs is clearly most unlikely to be random. When the curve $y = 4x - 0.1x^2$ is fitted to the data, the order of the signs of the residuals is: $+\,-\,-\,+\,+\,-\,0\,-\,+\,-\,+$, and the sum of squares is only 1.29. There is thus every evidence that the points are better fitted by the curve than by the straight line. For the $y$ values 5, 16, and 27 the straight line predicts concentrations of 1.15, 4.83, and 8.51 units respectively, while the curve predicts concentractions of 1.29, 4.51, and 8.60 units respectively. In the light of the analysis of the residual, the latter values are certainly more reliable. It must not be overlooked, however, that other curves might fit the experimental points even better than the one suggested here.

The example also shows that, in cases where there is a reasonable number of calibration points, and where the random errors are not very large, treating the curve as a series of short straight segments is an attractive approximate method. If ten such linear segments are drawn, i.e. if all the calibration points are used, the concentrations calculated by linear interpolation for signals of 5, 16, and 27 units are 1.36, 4.50 and 8.65 units respectively. Even if only five segments are drawn, i.e. only the points $x = 0, 2, 4$ etc. are used, the predicted concentrations are 1.32, 4.37, and 8.51 units. In view of the ease of calculation and the relatively small errors compared with the values derived from the fitted curve, the attractions of this method are apparent.

# CHAPTER 5

1. The mean is 9.96 ml, and the median (the average of the second and third ranked measurements) is 9.90 ml. The $Q$-test shows that the value 10.20 cannot quite be rejected at $P = 0.05$. If it *were* rejected the mean would move to 9.88 ml and the median to 9.89 ml: the median is insensitive to the presence of outliers.

2. (a) Use of the sign test: compared with the median, the experimental values give signs of $-\,+\,0\,+\,-\,+\,+\,+\,+$. In effect we thus have 8 readings, 6 with $+$ signs. The chance of six signs out of 8 being the same is 0.29, much greater than 0.05. So we retain the null hypothesis that the data could have come from a symmetrical population with median sulphur content of 0.10%.

(b) In the signed rank test, the zero value is again neglected. The remaining data compared with the median give ranked differences of $-0.01, 0.01, 0.01, -0.02, 0.02, 0.02, 0.04, 0.07$. The signed ranks, taking into account the ties, are thus $-2, 2, 2, -5, 5, 5, 7, 8$. The negative ranks total $(-)7$ but, at $P = 0.05$, the critical value for $n = 8$ is only 3. Again, therefore, the null hypothesis is retained.

(c) To use a range test, we calculate the mean ($\bar{x}$), which is 0.1156, and the range (w), 0.09. Then $T_I$ is given by $|\bar{x} - V|/w = 0.173$. The critical value in this case is 0.255, so for the third time the null hypothesis is retained. Contrary to superficial appearances, perhaps, the sample of oils cannot be said to fail the specification.

3. This problem can also be solved by using the sign test and the signed rank test. If the e.i.d. values are subtracted from r.i.d. values, the signs of the differences are $+ - + + + + + + 0 +$. In effect we have 9 results, with 8 + signs and one $-$ sign. The probability of 8 signs out of nine being the same is only 0.04, so at the significance level $P = 0.05$ we can reject the null hypothesis: the results *are* significantly different. In the signed rank test, the magnitude of the negative rank is only 2.5. This is well below the critical value of 8, so again the null hypothesis is rejected, and the significant difference between the two sets of results is confirmed. Note, however, that the results of the signed rank test should be interpreted with caution if there are a lot of tied ranks, as there are both in this question and the previous one.

4. The new method gives a mean of 64.7 mg/100ml and a range of 2.1 mg/100 ml: corresponding figures for the standard method are 65.86 and 0.9 mg/100 ml. From these figures, $T_d$ is easily shown to be 0.773, greater than the critical value of 0.61, so the means *do* differ significantly. The range ratio $F_R$ is 2.33, compared with the critical value of 3.2, so the precisions of the methods are not significantly different. (The $F$-test gives the same result).

5. If the values are arranged in ascending order, the median is found to be 23.5. The individual values, in order of use of the carbon rods, thus have signs $+ + +$ $- - - - - + +$. There are thus three runs. But for $N = M = 5$ the critical value is 2, so the given sequence might well have been a random one.

6. This problem is easily solved by the $U$-test. If the "beer" values are expected to be greater than the "lager" values, the number of lager values that are greater than the individual beer values should be small. It is actually 4.5, counting 0.5 for the one tie that occurs. The critical value in a *one-tailed* test is 5, so we can (just) reject the null hypothesis at the $P = 0.05$ level and say that one particular lager does produce lower blood alcohol levels than one particular beer!

7. The ranks are as follows:

| Instrument: | A | B | C | D | E | F | G |
|---|---|---|---|---|---|---|---|
| Student ranking: | 3 | 1 | 5 | 4 | 7 | 6 | 2 |
| Staff ranking: | 5 | 3 | 6 | 2 | 4 | 7 | 1 |
| $d$: | 2 | 2 | 1 | 2 | 3 | 1 | 1 |
| $d^2$: | 4 | 4 | 1 | 4 | 9 | 1 | 1 |

Thus $\rho$ is $1 - (6 \times 24)/(7 \times 48) = 0.571$. This is well below the critical value of 0.786 ($P = 0.05$ and $n = 7$), so the students and staff do not seem to agree – the correlation is not significant.

8. If the $x$ values are the distances, and the $y$ values the mercury levels, then Theil's method gives $a = 2.575$, and $b = -0.125$. These values are astonishingly close to the (parametric) 'least squares' values of 2.573 and $-0.122$.

9. To test $\bar{x} = 1.0$ and $\sigma = 0.2$, we write $z = (x - 1.0)/0.2$, and obtain 1.5, 2.5, $-1.5$, $-0.5$, 0, 0.5, $-1.0$, 4.0, $-3.0$, and 1.5. Plotted against the cumulative distribution function for the standard normal distribution, these values give a trace that is fairly symmetrical, but much broader than the standard curve, indicating that the assumed mean is not very far wrong, but the estimated standard deviation is far too low. The maximum difference is about $+0.335$ at $z = 1.5$. This is higher than the critical value of 0.262, confirming that the null hypothesis must be rejected, and that the values do not fit this particular normal distribution. To test for another normal distribution, we calculate 1.08 and 0.41 as our estimates of the mean and standard deviation respectively. These give $z$ values of 0.54, 1.02, $-0.93$, $-0.44$, $-0.20$, 0.05, $-0.68$, 1.76, $-1.66$, and 0.54. The maximum difference in this case is only ca. 0.11 at $z = 0.54$, so the null hypothesis is retained; the data fit this normal distribution very well.

## CHAPTER 6

1. An example of two-way ANOVA with replication.

| Source of variation | Sum of squares | D.f. | Mean square |
|---|---|---|---|
| Between row | 5.06778 | 2 | 2.53 |
| Between column | 0.18778 | 2 | 0.0939 |
| Interaction | 0.10222 | 4 | 0.0256 |
| Residual | 0.36500 | 9 | 0.0406 |
| Total | 5.72278 | 17 | |

The residual mean square is greater than the interaction mean square so there is no significant labroatory-sample interaction. To test whether there is a significant difference between laboratories the between column mean square is compared with the residual mean square, giving:

$$F_{2,9} = 0.0939/0.0406 = 2.31$$

The critical value of $F_{2,9}(P = 0.05)$ is 4.256, so the calculated value is not significant.

2. Vertex 1 should be rejected in forming the new simplex.

|  | Factors | | | | | |
| --- | --- | --- | --- | --- | --- | --- |
|  | A | B | C | D | E | Response |
| Vertex 2 | 6.0 | 4.3 | 9.5 | 6.9 | 6.0 | 8 |
| Vertex 3 | 2.5 | 11.5 | 9.5 | 6.9 | 6.0 | 10 |
| Vertex 5 | 2.5 | 4.3 | 9.5 | 9.7 | 6.0 | 11 |
| Vertex 6 | 2.5 | 4.3 | 9.5 | 6.9 | 9.6 | 9 |
| Vertex 7 | 3.3 | 6.7 | 12.5 | 7.7 | 7.0 | 12 |
| Sum | 16.8 | 31.1 | 50.5 | 38.1 | 34.6 | |
| Sum $\div n$ | 3.4 | 6.2 | 10.1 | 7.6 | 6.9 | |
| Rejected vertex (i.e. 1) | 1.0 | 3.0 | 2.0 | 6.0 | 5.0 | |
| Displacement | 2.4 | 3.2 | 8.1 | 1.6 | 1.9 | |
| Vertex 8 | 5.8 | 9.4 | 18.1 | 9.2 | 8.8 | 7 |

(All values in the above table are given to 1 decimal place).

3. An example of two-way ANOVA without replication.

| Source of variation | Sum of squares | Degrees of freedom | Mean square |
| --- | --- | --- | --- |
| Between row | 146.25 | 3 | 48.8 |
| Between column | 110.16667 | 2 | 55.1 |
| Residual | 254.5 | 6 | 42.4 |
| Total | 510.91667 | 11 | |

(a) Comparing the between solution (i.e. row) mean square with the residual mean square gives $F_{3,6} = 1.15$, which is not significant ($P = 0.05$).
(b) Comparing the between method (i.e. column) mean square gives $F_{2,6} = 1.30$, which is not significant ($P = 0.05$).

4. An example of two-way ANOVA with replication.

| Source of variation | Sum of squares | Degrees of freedom | Mean square |
| --- | --- | --- | --- |
| Between row | 20.8768749 | 3 | 6.959 |
| Between column | 0.14062595 | 1 | 0.141 |
| Interaction | 0.11187458 | 3 | 0.037 |
| Residual | 9.92499971 | 8 | 1.241 |
| Total | 31.0543752 | 15 | |

There is no significant interaction or difference between methods since in both cases the corresponding mean square is less than the residual mean square.

5. Using the following notation:

$G$ = purge gas (Ar − low, Ar + $O_2$ − high)
$T$ = purge time (20 sec − low, 60 sec − high)
$Q$ = quartz cell temperature (900°C − low, 1000°C − high)

the calculated effects are:

| Single factor | Effect |
|---|---|
| G | 18.25 |
| T | −12.25 |
| Q | 55.25 |

| Two-factor | |
|---|---|
| GT | 13.75 |
| TQ | −5.25 |
| GQ | −0.75 |

| Three-factor | |
|---|---|
| GTQ | 3.75 |

6.
| Single factor | Effect |
|---|---|
| A | 0.215 |
| C | 0.005 |
| T | 0.265 |

| Two-factor | |
|---|---|
| AC | 0.005 |
| CT | 0.025 |
| AT | 0.065 |

Three-factor, ATC    0.005

7. Using the following notation:

$N$ = nitric acid concentration ($0.1M$ – low, $1.0M$ – high)
$L$ = lanthanum concentration ($0M$ – low, $0.1M$ – high)

the calculated effects are:

| Single factor | Effect | Sum of squares | Degrees of freedom |
|---|---|---|---|
| N | 9.5 | 180.5 | 1 |
| L | 22.5 | 1012.5 | 1 |
| Interaction | | | |
| NL | 3.5 | 24.5 | 1 |

To test for significant interaction the interaction mean square is compared with the residual mean square, giving $F_{1,4} = 24.5/4 = 6.125$. The critical value of $F$ is 7.709 ($P = 0.05$), so there is no significant interaction.

8. The variance of the mean is 4 in both cases. The second scheme is more economical when the ratio of cost of sampling to cost of analysis exceeds 2:1.

# Appendices

---

## APPENDIX 1  SUMMARY OF STATISTICAL TESTS

The following tables summarize the statistical tests described in this book. The tests are grouped according to their application, with brief details and comments in each case, and a reference to the page on which further details may be found. It is hoped that this will help the reader in deciding on the statistical test appropriate to a particular problem.

(1) Tests on averages

| Test name or test statistic | Used to test whether | Refer to page | Comments |
|---|---|---|---|
| $t$ | (a) a sample mean differs from a standard value<br>(b) the means of two samples differ<br>(c) paired samples differ | 52<br>53<br>56 | Assumes that populations are normal |
| Sign test | (a) a sample mean differs from a standard value<br>(b) paired samples differ | 121<br>122 | Non-parametric |
| $T_{\mathrm{I}}$ | (a) a sample mean differs from a standard value | 124 | Assumes that population is normal; quicker than $t$-test |
| $T_{\mathrm{d}}$ | (a) the means of two samples differ | 125 | Assumes that populations are normal; quicker than $t$-test |
| $L$ (Lord's range test) | (a) the means of two samples differ | 125 | Assumes that populations are normal; quicker than $t$-test |
| Wilcoxon signed rank test | (a) a sample mean differs from a standard value<br>(b) paired samples differ | 126<br>127 | Non-parametric but assumes that the population is symmetrical; requires $n \geq 5$ for one-tailed and $n \geq 6$ for two-tailed tests |
| Wilcoxon rank sum test | (a) the means of two samples differ | 129 | Non-parametric; minimum sample sizes: one-tailed test, $n_1 = n_2 = 3$; two-tailed test, $n_1 = 3, n_2 = 5$ |
| Mann–Whitney $U$-test | (a) the means of two samples differ | 130 | Simpler version of Wilcoxon rank sum test |

(2) Tests on spread

| Test name or test statistic | Used to test whether: | Refer to page | Comments |
| --- | --- | --- | --- |
| $F$ | variances of two samples differ | 57 | Assumes normality |
| $F_R$ | ranges of two samples differ | 125 | Simpler to perform than $F$-test; assumes normality |

(3) Tests on runs

| | | | |
| --- | --- | --- | --- |
| Wald–Wolfowitz | a sequence is random | 123 | For minimum numbers of data points see Table A.8 |

(4) Tests on correlation

| | | | |
| --- | --- | --- | --- |
| Product–moment correlation coefficient, $r$ | data are linearly correlated | 85 | Assumes that errors are normally distributed |
| Spearman rank correlation coefficient, $\rho$ | data are linearly correlated | 131 | Simpler to calculate than $r$; non-parametric |

(5) Goodness-of-fit tests

| | | | |
| --- | --- | --- | --- |
| $\chi^2$ (chi-squared) | observed frequencies differ from expected frequencies | 74 | Requires total sample $\geqslant 50$ and individual frequencies $\geqslant 5$ |
| Kolmogorov–Smirnov | data conform to a specified distribution | 134 | Can be used for small samples; particularly useful in testing for normality |

## APPENDIX 2   STATISTICAL TABLES

The following tables are presented for the convenience of the reader, and for use with the simple statistical tests, examples and exercises in this book. They are presented in a format that is compatible with the needs of analytical chemists: the significance level $P = 0.05$ has been used in most cases, and it has been assumed that the number of measurements available is fairly small. Except where stated otherwise, these abbreviated tables have been taken, with permission, from *Elementary Statistics Tables* by Henry R. Neave, published by George Allen & Unwin Ltd. (Table A.1-A.3, A.5-A.7, and A.12-A.16). The reader requiring statistical data corresponding to significance levels and/or numbers of measurements not covered in the tables is referred to the sources.

**Table A.1** – The $t$-distribution

| | 95% | 98% | 99% |
|---|---|---|---|
| Value of $t$ for a confidence interval of | 95% | 98% | 99% |
| Critical value of $|t|$ for $P$ values of | 0.05 | 0.02 | 0.01 |
| Number of degrees of freedom | | | |
| 1 | 12.71 | 31.82 | 63.66 |
| 2 | 4.30 | 6.96 | 9.92 |
| 3 | 3.18 | 4.54 | 5.84 |
| 4 | 2.78 | 3.75 | 4.60 |
| 5 | 2.57 | 3.36 | 4.03 |
| 6 | 2.45 | 3.14 | 3.71 |
| 7 | 2.36 | 3.00 | 3.50 |
| 8 | 2.31 | 2.90 | 3.36 |
| 9 | 2.26 | 2.82 | 3.25 |
| 10 | 2.23 | 2.76 | 3.17 |
| 12 | 2.18 | 2.68 | 3.05 |
| 14 | 2.14 | 2.62 | 2.98 |
| 16 | 2.12 | 2.58 | 2.92 |
| 18 | 2.10 | 2.55 | 2.88 |
| 20 | 2.09 | 2.53 | 2.85 |
| 30 | 2.04 | 2.46 | 2.75 |
| 50 | 2.01 | 2.40 | 2.68 |
| ∞ | 1.96 | 2.33 | 2.58 |

**Table A.2** – Critical values of $F$ for a one-tailed test ($P = 0.05$)

| $\nu_1$ | 1 | 2 | 3 | 4 | 5 | 6 | 7 | 8 | 9 | 10 | 12 | 15 | 20 |
|---|---|---|---|---|---|---|---|---|---|---|---|---|---|
| $\nu_2$ | | | | | | | | | | | | | |
| 1 | 161.4 | 199.5 | 215.7 | 224.6 | 230.2 | 234.0 | 236.8 | 238.9 | 240.5 | 241.9 | 243.9 | 245.9 | 248.0 |
| 2 | 18.51 | 19.00 | 19.16 | 19.25 | 19.30 | 19.33 | 19.35 | 19.37 | 19.38 | 19.40 | 19.41 | 19.43 | 19.45 |
| 3 | 10.13 | 9.552 | 9.277 | 9.117 | 9.013 | 8.941 | 8.887 | 8.845 | 8.812 | 8.786 | 8.745 | 8.703 | 8.660 |
| 4 | 7.709 | 6.944 | 6.591 | 6.388 | 6.256 | 6.163 | 6.094 | 6.041 | 5.999 | 5.964 | 5.912 | 5.858 | 5.803 |
| 5 | 6.608 | 5.786 | 5.409 | 5.192 | 5.050 | 4.950 | 4.876 | 4.818 | 4.772 | 4.735 | 4.678 | 4.619 | 4.558 |
| 6 | 5.987 | 5.143 | 4.757 | 4.534 | 4.387 | 4.284 | 4.207 | 4.147 | 4.099 | 4.060 | 4.000 | 3.938 | 3.874 |
| 7 | 5.591 | 4.737 | 4.347 | 4.120 | 3.972 | 3.866 | 3.787 | 3.726 | 3.677 | 3.637 | 3.575 | 3.511 | 3.445 |
| 8 | 5.318 | 4.459 | 4.066 | 3.838 | 3.687 | 3.581 | 3.500 | 3.438 | 3.388 | 3.347 | 3.284 | 3.218 | 3.150 |
| 9 | 5.117 | 4.256 | 3.863 | 3.633 | 3.482 | 3.374 | 3.293 | 3.230 | 3.179 | 3.137 | 3.073 | 3.006 | 2.936 |
| 10 | 4.965 | 4.103 | 3.708 | 3.478 | 3.326 | 3.217 | 3.135 | 3.072 | 3.020 | 2.978 | 2.913 | 2.845 | 2.774 |
| 11 | 4.844 | 3.982 | 3.587 | 3.357 | 3.204 | 3.095 | 3.012 | 2.948 | 2.896 | 2.854 | 2.788 | 2.719 | 2.646 |
| 12 | 4.747 | 3.885 | 3.490 | 3.259 | 3.106 | 2.996 | 2.913 | 2.849 | 2.796 | 2.753 | 2.687 | 2.617 | 2.544 |
| 13 | 4.667 | 3.806 | 3.411 | 3.179 | 3.025 | 2.915 | 2.832 | 2.767 | 2.714 | 2.671 | 2.604 | 2.533 | 2.459 |
| 14 | 4.600 | 3.739 | 3.344 | 3.112 | 2.958 | 2.848 | 2.764 | 2.699 | 2.646 | 2.602 | 2.534 | 2.463 | 2.388 |
| 15 | 4.543 | 3.682 | 3.287 | 3.056 | 2.901 | 2.790 | 2.707 | 2.641 | 2.588 | 2.544 | 2.475 | 2.403 | 2.328 |
| 16 | 4.494 | 3.634 | 3.239 | 3.007 | 2.852 | 2.741 | 2.657 | 2.591 | 2.538 | 2.494 | 2.425 | 2.352 | 2.276 |
| 17 | 4.451 | 3.592 | 3.197 | 2.965 | 2.810 | 2.699 | 3.614 | 2.548 | 2.494 | 2.450 | 2.381 | 2.308 | 2.230 |
| 18 | 4.414 | 3.555 | 3.160 | 2.928 | 2.773 | 2.661 | 2.577 | 2.510 | 2.456 | 2.412 | 2.342 | 2.269 | 2.191 |
| 19 | 4.381 | 3.522 | 3.127 | 2.895 | 2.740 | 2.628 | 2.544 | 2.477 | 2.423 | 2.378 | 2.308 | 2.234 | 2.155 |
| 20 | 4.351 | 3.493 | 3.098 | 2.866 | 2.711 | 2.599 | 2.514 | 2.447 | 2.393 | 2.348 | 2.278 | 2.203 | 2.124 |

$\nu_1$ = number of degrees of freedom of the numerator and $\nu_2$ = number of degrees of freedom of the denominator.

**Table A.3** – Critical values of $F$ for a two-tailed test ($P = 0.05$)

| $\nu_1$ | 1 | 2 | 3 | 4 | 5 | 6 | 7 | 8 | 9 | 10 | 12 | 15 | 20 |
|---|---|---|---|---|---|---|---|---|---|---|---|---|---|
| $\nu_2$ | | | | | | | | | | | | | |
| 1 | 647.8 | 799.5 | 864.2 | 899.6 | 921.8 | 937.1 | 948.2 | 956.7 | 963.3 | 968.6 | 976.7 | 984.9 | 993.1 |
| 2 | 38.51 | 39.00 | 39.17 | 39.25 | 39.30 | 39.33 | 39.36 | 39.37 | 39.39 | 39.40 | 39.41 | 39.43 | 39.45 |
| 3 | 17.44 | 16.04 | 15.44 | 15.10 | 14.88 | 14.73 | 14.62 | 14.54 | 14.47 | 14.42 | 14.34 | 14.25 | 14.17 |
| 4 | 12.22 | 10.65 | 9.979 | 9.605 | 9.364 | 9.197 | 9.074 | 8.980 | 8.905 | 8.844 | 8.751 | 8.657 | 8.560 |
| 5 | 10.01 | 8.434 | 7.764 | 7.388 | 7.146 | 6.978 | 6.853 | 6.757 | 6.681 | 6.619 | 6.525 | 6.428 | 6.329 |
| 6 | 8.813 | 7.260 | 6.599 | 6.227 | 5.988 | 5.820 | 5.695 | 5.600 | 5.523 | 5.461 | 5.366 | 5.269 | 5.168 |
| 7 | 8.073 | 6.542 | 5.890 | 5.523 | 5.285 | 5.119 | 4.995 | 4.899 | 4.823 | 4.761 | 4.666 | 4.568 | 4.467 |
| 8 | 7.571 | 6.059 | 5.416 | 5.053 | 4.817 | 4.652 | 4.529 | 4.433 | 4.357 | 4.295 | 4.200 | 4.101 | 3.999 |
| 9 | 7.209 | 5.715 | 5.078 | 4.718 | 4.484 | 4.320 | 4.197 | 4.102 | 4.026 | 3.964 | 3.868 | 3.769 | 3.667 |
| 10 | 6.937 | 5.456 | 4.826 | 4.468 | 4.236 | 4.072 | 3.950 | 3.855 | 3.779 | 3.717 | 3.621 | 3.522 | 3.419 |
| 11 | 6.724 | 5.256 | 4.630 | 4.275 | 4.044 | 3.881 | 3.759 | 3.664 | 3.588 | 3.526 | 3.430 | 3.330 | 3.226 |
| 12 | 6.554 | 5.096 | 4.474 | 4.121 | 3.891 | 3.728 | 3.607 | 3.512 | 3.436 | 3.374 | 3.277 | 3.177 | 3.073 |
| 13 | 6.414 | 4.965 | 4.347 | 3.996 | 3.767 | 3.604 | 3.483 | 3.388 | 3.312 | 3.250 | 3.153 | 3.053 | 2.948 |
| 14 | 6.298 | 4.857 | 4.242 | 3.892 | 3.663 | 3.501 | 3.380 | 3.285 | 2.209 | 3.147 | 3.050 | 2.949 | 2.844 |
| 15 | 6.200 | 4.765 | 4.153 | 3.804 | 3.576 | 3.415 | 3.293 | 3.199 | 3.123 | 3.060 | 2.963 | 2.862 | 2.756 |
| 16 | 6.115 | 4.687 | 4.077 | 3.729 | 3.502 | 3.341 | 3.219 | 3.125 | 3.049 | 2.986 | 2.889 | 2.788 | 2.681 |
| 17 | 6.042 | 4.619 | 4.011 | 3.665 | 3.438 | 3.277 | 3.156 | 3.061 | 2.985 | 2.922 | 2.825 | 2.723 | 2.616 |
| 18 | 5.978 | 4.560 | 3.954 | 3.608 | 3.382 | 3.221 | 3.100 | 3.005 | 2.929 | 2.866 | 2.769 | 2.667 | 2.559 |
| 19 | 5.922 | 4.508 | 3.903 | 3.559 | 3.333 | 3.172 | 3.051 | 2.956 | 2.880 | 2.817 | 2.720 | 2.617 | 2.509 |
| 20 | 5.871 | 4.461 | 3.859 | 3.515 | 3.289 | 3.128 | 3.007 | 2.913 | 2.837 | 2.774 | 2.676 | 2.573 | 2.464 |

$\nu_1$ = number of degrees of freedom of the numerator and $\nu_2$ = number of degrees of freedom of the denominator.

**Table A.4** – Critical values of $Q$ $(P = 0.05)$

| Sample size | Critical value |
|---|---|
| 4 | 0.831 |
| 5 | 0.717 |
| 6 | 0.621 |
| 7 | 0.570 |
| 8 | 0.524 |
| 9 | 0.492 |
| 10 | 0.464 |

Taken from E. P. King, *J. Am. Statist. Assoc.*, 1958, **48**, 531, by permission of the American Statistical Association.

**Table A.5** – Critical values of $\chi^2$ $(P = 0.05)$

| Number of degrees of freedom | Critical value |
|---|---|
| 1 | 3.84 |
| 2 | 5.99 |
| 3 | 7.81 |
| 4 | 9.49 |
| 5 | 11.07 |
| 6 | 12.59 |
| 7 | 14.07 |
| 8 | 15.51 |
| 9 | 16.92 |
| 10 | 18.31 |

**Table A.6** – Random numbers

| | | | | | | | | | |
|---|---|---|---|---|---|---|---|---|---|
| 02484 | 88139 | 31788 | 35873 | 63259 | 99886 | 20644 | 41853 | 41915 | 02944 |
| 83680 | 56131 | 12238 | 68291 | 95093 | 07362 | 74354 | 13071 | 77901 | 63058 |
| 37336 | 63266 | 18632 | 79781 | 09184 | 83909 | 77232 | 57571 | 25413 | 82680 |
| 04060 | 46030 | 23751 | 61880 | 40119 | 88098 | 75956 | 85250 | 05015 | 99184 |
| 62040 | 01812 | 46847 | 79352 | 42478 | 71784 | 65864 | 84904 | 48901 | 17115 |
| 96417 | 63336 | 88491 | 73259 | 21086 | 51932 | 32304 | 45021 | 61697 | 73953 |
| 42293 | 29755 | 24119 | 62125 | 33717 | 20284 | 55606 | 33308 | 51007 | 68272 |
| 31378 | 35714 | 00941 | 53042 | 99174 | 30596 | 67769 | 59343 | 53193 | 19203 |
| 27098 | 38959 | 49721 | 69341 | 40475 | 55998 | 87510 | 55523 | 15549 | 32402 |
| 66527 | 73898 | 66912 | 76300 | 52782 | 29356 | 35332 | 52387 | 29194 | 21591 |
| 61621 | 52967 | 40644 | 91293 | 80576 | 67485 | 88715 | 45293 | 59454 | 76218 |
| 18798 | 99633 | 32948 | 49802 | 40261 | 35555 | 76229 | 00486 | 64236 | 74782 |
| 36864 | 66460 | 87303 | 13788 | 04806 | 31140 | 75253 | 79692 | 47618 | 20024 |
| 10346 | 28822 | 51891 | 04097 | 98009 | 58042 | 67833 | 23539 | 37668 | 16324 |
| 20582 | 49576 | 91822 | 63807 | 99450 | 18240 | 70002 | 75386 | 26035 | 21459 |
| 12023 | 82328 | 54810 | 64766 | 58954 | 76201 | 78456 | 98467 | 34166 | 84186 |
| 48255 | 20815 | 51322 | 04936 | 33413 | 43128 | 21643 | 90674 | 98858 | 26060 |
| 92956 | 09401 | 58892 | 59686 | 10899 | 89780 | 57080 | 82799 | 70178 | 40399 |
| 87300 | 04729 | 57966 | 95672 | 49036 | 24993 | 69827 | 67637 | 09472 | 63356 |
| 69101 | 21192 | 00256 | 81645 | 48500 | 73237 | 95420 | 98974 | 36036 | 21781 |
| 22084 | 03117 | 96937 | 86176 | 80102 | 48211 | 61149 | 71246 | 19993 | 79708 |
| 28000 | 44301 | 40028 | 88132 | 07083 | 50818 | 09104 | 92449 | 27860 | 90196 |
| 41662 | 20930 | 32856 | 91566 | 64917 | 18709 | 79884 | 44742 | 18010 | 11599 |
| 91398 | 16841 | 51399 | 82654 | 00857 | 21068 | 94121 | 39197 | 27752 | 67308 |
| 46560 | 00597 | 84561 | 42334 | 06695 | 26306 | 16832 | 63140 | 13762 | 15598 |

### Table A.7 – The sign test

The table uses the binomial distribution with $p = 0.5$ to give the probabilities of $r$ or less successes for $n = 4$-15. These values correspond to a one-tailed sign test and should be doubled for a two-tailed test.

| n | $r = 0$ | 1 | 2 | 3 | 4 | 5 | 6 | 7 |
|----|-------|-------|-------|-------|-------|-------|-------|-------|
| 4  | 0.063 | 0.313 | 0.688 |       |       |       |       |       |
| 5  | 0.031 | 0.188 | 0.500 |       |       |       |       |       |
| 6  | 0.016 | 0.109 | 0.344 | 0.656 |       |       |       |       |
| 7  | 0.008 | 0.063 | 0.227 | 0.500 |       |       |       |       |
| 8  | 0.004 | 0.035 | 0.144 | 0.363 | 0.637 |       |       |       |
| 9  | 0.002 | 0.020 | 0.090 | 0.254 | 0.500 |       |       |       |
| 10 | 0.001 | 0.011 | 0.055 | 0.172 | 0.377 | 0.623 |       |       |
| 11 | 0.001 | 0.006 | 0.033 | 0.113 | 0.274 | 0.500 |       |       |
| 12 | 0.000 | 0.003 | 0.019 | 0.073 | 0.194 | 0.387 | 0.613 |       |
| 13 | 0.000 | 0.002 | 0.011 | 0.046 | 0.133 | 0.290 | 0.500 |       |
| 14 | 0.000 | 0.001 | 0.006 | 0.029 | 0.090 | 0.212 | 0.395 | 0.605 |
| 15 | 0.000 | 0.000 | 0.004 | 0.018 | 0.059 | 0.151 | 0.304 | 0.500 |

### Table A.8 – The Wald-Wolfowitz runs test

| N | M | At $P = 0.05$, the number of runs is significant if it is: | |
|---|---|---|---|
| | | Less than | Greater than |
| 2 | 12–20 | 3 | NA |
| 3 | 6–14 | 3 | NA |
| 3 | 15–20 | 4 | NA |
| 4 | 5–6 | 3 | 8 |
| 4 | 7 | 3 | NA |
| 4 | 8–15 | 4 | NA |
| 4 | 16–20 | 5 | NA |
| 5 | 5 | 3 | 9 |
| 5 | 6 | 4 | 9 |
| 5 | 7–8 | 4 | 10 |
| 5 | 9–10 | 4 | NA |
| 5 | 11–17 | 5 | NA |
| 6 | 6 | 4 | 10 |
| 6 | 7–8 | 4 | 11 |
| 6 | 9–12 | 5 | 12 |
| 6 | 13–18 | 6 | NA |
| 7 | 7 | 4 | 12 |
| 7 | 8 | 5 | 12 |
| 7 | 9 | 5 | 13 |
| 7 | 10–12 | 6 | 13 |
| 8 | 8 | 5 | 13 |
| 8 | 9 | 6 | 13 |
| 8 | 10–11 | 6 | 14 |
| 8 | 12–15 | 7 | 15 |

Adapted from F. Swed and C. Eisenhart, *Ann. Math. Statist.*, 1943, **83,** by permission of the Institute of Mathematical Statistics. The test cannot be applied to data with $N$, $M$ smaller than the given numbers, or to cases marked NA.

## Table A.9 – The $T_I$ test

Critical values for one-tailed and two-tailed tests at $P = 0.05$.

| $n$ | One-tailed test | Two-tailed test |
|---|---|---|
| 2 | 3.175 | 6.353 |
| 3 | 0.885 | 1.304 |
| 4 | 0.529 | 0.717 |
| 5 | 0.388 | 0.507 |
| 6 | 0.312 | 0.399 |
| 7 | 0.263 | 0.333 |
| 8 | 0.230 | 0.288 |
| 9 | 0.205 | 0.255 |
| 10 | 0.186 | 0.230 |

## Table A.10 – The $T_d$ Test and Lord's range test

Critical values for $T_d$ and $L$ at $P = 0.05$

| $n_1 = n_2$ | $T_d$ | $L$ |
|---|---|---|
| 2 | 3.43 | 1.71 |
| 3 | 1.27 | 0.64 |
| 4 | 0.81 | 0.41 |
| 5 | 0.61 | 0.31 |
| 6 | 0.50 | 0.25 |
| 7 | 0.43 | 0.21 |
| 8 | 0.37 | 0.19 |
| 9 | 0.33 | 0.17 |
| 10 | 0.30 | 0.15 |

Table A.9 and A.10 are taken from E. Lord, Biometrika, 1947, **34,** 66, by permission of the Biometrika Trustees.

## Table A.11 – The substitute $F$ ($F_R$) test

Critical $F_R$ values for one-tailed and two-tailed tests at $P = 0.05$.

| Number of measurements in numerator and denominator | One-tailed test | Two-tailed test |
|---|---|---|
| 2 | 12.7 | 25.5 |
| 3 | 4.4 | 6.3 |
| 4 | 3.1 | 4.0 |
| 5 | 2.6 | 3.2 |
| 6 | 2.3 | 2.8 |
| 7 | 2.1 | 2.5 |
| 8 | 2.0 | 2.3 |
| 9 | 1.9 | 2.2 |
| 10 | 1.9 | 2.1 |

Adapted from F. R. Link, *Ann. Math. Statist.*, 1949, **20,** 257 by permission of the Institute of Mathematical Statistics.

**Table A.12** – Wilcoxon signed rank test

Critical values for the test statistic at $P = 0.05$.

| $n$ | One-tailed test | Two-tailed test |
|---|---|---|
| 5 | 0 | NA |
| 6 | 2 | 0 |
| 7 | 3 | 2 |
| 8 | 5 | 3 |
| 9 | 8 | 5 |
| 10 | 10 | 8 |
| 11 | 13 | 10 |
| 12 | 17 | 13 |
| 13 | 21 | 17 |
| 14 | 25 | 21 |
| 15 | 30 | 25 |

The null hypothesis can be rejected when the test statistic is $\leqslant$ the tabulated value. NA indicates that the test cannot be applied.

**Table A.13** – Wilcoxon rank sum test; Mann–Whitney $U$-test

Critical values for $U$ or the lower of $T_1$ and $T_2$ at $P = 0.05$

| $n_1$ | $n_2$ | One-tailed test | Two-tailed test |
|---|---|---|---|
| 3 | 3 | 0 | NA |
| 3 | 4 | 0 | NA |
| 3 | 5 | 1 | 0 |
| 3 | 6 | 2 | 1 |
| 4 | 4 | 1 | 0 |
| 4 | 5 | 2 | 1 |
| 4 | 6 | 3 | 2 |
| 4 | 7 | 4 | 3 |
| 5 | 5 | 4 | 2 |
| 5 | 6 | 5 | 3 |
| 5 | 7 | 6 | 5 |
| 6 | 6 | 7 | 5 |
| 6 | 7 | 8 | 6 |
| 7 | 7 | 11 | 8 |

The null hypothesis can be rejected when $U$ or the lower $T$ value is $\leqslant$ the tabulated value. NA indicates that the test cannot be applied.

**Table A.14** – The Spearman rank correlation coefficient

Critical values for $\rho$ at $P = 0.05$

| $n$ | One-tailed test | Two-tailed test |
|---|---|---|
| 5 | 0.900 | 1.000 |
| 6 | 0.829 | 0.886 |
| 7 | 0.714 | 0.786 |
| 8 | 0.643 | 0.738 |
| 9 | 0.600 | 0.700 |
| 10 | 0.564 | 0.649 |
| 11 | 0.536 | 0.618 |
| 12 | 0.504 | 0.587 |
| 13 | 0.483 | 0.560 |
| 14 | 0.464 | 0.538 |
| 15 | 0.446 | 0.521 |
| 16 | 0.429 | 0.503 |
| 17 | 0.414 | 0.488 |
| 18 | 0.401 | 0.472 |
| 19 | 0.391 | 0.460 |
| 20 | 0.380 | 0.447 |

**Table A.15** – The Kolmogorov goodness-of-fit test

Critical values for one-tailed and two-tailed tests at $P = 0.05$. The appropriate value is compared with the maximum difference between the experimental and theoretical cumulative frequency curves, as described in the text.

| $n$ | One-tailed test | Two-tailed test |
|---|---|---|
| 1 | 0.950 | 0.975 |
| 2 | 0.776 | 0.842 |
| 3 | 0.636 | 0.708 |
| 4 | 0.565 | 0.624 |
| 5 | 0.509 | 0.563 |
| 6 | 0.468 | 0.519 |
| 7 | 0.436 | 0.483 |
| 8 | 0.410 | 0.454 |
| 9 | 0.388 | 0.430 |
| 10 | 0.369 | 0.409 |
| 11 | 0.352 | 0.392 |
| 12 | 0.338 | 0.375 |
| 13 | 0.326 | 0.361 |
| 14 | 0.314 | 0.349 |
| 15 | 0.304 | 0.338 |
| 16 | 0.295 | 0.327 |
| 17 | 0.286 | 0.318 |
| 18 | 0.278 | 0.309 |
| 19 | 0.271 | 0.301 |
| 20 | 0.265 | 0.294 |

**Table A.16** – The Kolmogorov test for normality

Critical values for one-tailed and two-tailed tests at $P = 0.05$. The appropriate value is compared with the maximum difference between the experimental and theoretical cumulative frequency curves, as described in the text.

| $n$ | One-tailed test | Two-tailed test |
|---|---|---|
| 3 | 0.367 | 0.376 |
| 4 | 0.345 | 0.375 |
| 5 | 0.319 | 0.343 |
| 6 | 0.297 | 0.323 |
| 7 | 0.280 | 0.304 |
| 8 | 0.265 | 0.288 |
| 9 | 0.252 | 0.274 |
| 10 | 0.241 | 0.262 |
| 11 | 0.231 | 0.251 |
| 12 | 0.222 | 0.242 |
| 13 | 0.215 | 0.234 |
| 14 | 0.208 | 0.226 |
| 15 | 0.201 | 0.219 |
| 16 | 0.195 | 0.213 |
| 17 | 0.190 | 0.207 |
| 18 | 0.185 | 0.202 |
| 19 | 0.181 | 0.197 |
| 20 | 0.176 | 0.192 |

# Index